T0328324

CONTROLLED RELEASE FERTILIZERS FOR SUSTAINABLE AGRICULTURE

CONTROLLED RELEASE FERTILIZERS FOR SUSTAINABLE AGRICULTURE

Edited by

F.B. LEWU

Department of Agriculture, Cape Peninsula University of Technology Wellington Campus, Wellington, South Africa

TATIANA VOLOVA

Department of Biotechnology, Siberian Federal University, Krasnoyarsk, Russia

SABU THOMAS

School of Energy Materials; School of Chemical Sciences, International and Inter-University Centre for Nanoscience and Nanotechnology, Mahatma Gandhi University, Kottayam, Kerala, India

RAKHIMOL K.R.

International and Inter-University Centre for Nanoscience and Nanotechnology, Mahatma Gandhi University, Kottayam, Kerala, India

ACADEMIC PRESS

An imprint of Elsevier

Notices
Knowledge and best practice in this field are constantly changing. As new research and experience broaden our understanding, changes in research methods, professional practices, or medical treatment may become necessary.

Practitioners and researchers must always rely on their own experience and knowledge in evaluating and using any information, methods, compounds, or experiments described herein. In using such information or methods they should be mindful of their own safety and the safety of others, including parties for whom they have a professional responsibility.

To the fullest extent of the law, neither the Publisher nor the authors, contributors, or editors, assume any liability for any injury and/or damage to persons or property as a matter of products liability, negligence or otherwise, or from any use or operation of any methods, products, instructions, or ideas contained in the material herein.

Library of Congress Cataloging-in-Publication Data
A catalog record for this book is available from the Library of Congress

British Library Cataloguing-in-Publication Data
A catalogue record for this book is available from the British Library

ISBN: 978-0-12-819555-0

For information on all Academic Press publications visit our website at https://www.elsevier.com/books-and-journals

Publisher: Charlotte Cockle
Acquisitions Editor: Nancy Maragioglio
Editorial Project Manager: Lena Sparks
Production Project Manager: Joy Christel Neumarin Honest Thangiah
Cover Designer: Matthew Limbert

Typeset by TNQ Technologies

Contents

Contributors

Ashitha A.
School of Biosciences, Mahatma Gandhi University, Kottayam, Kerala, India

Soumia Aboulhrouz
VARENA Center, MAScIR Foundation, Rabat Design, Rabat, Morocco

Aiman E. Al-Rawajfeh
Department of Chemical Engineering, Tafila Technical University, Tafila, Jordan

Mohammad R. Alrbaihat
Ministry of Education, Ajman, United Arab Emirates

Ehab M. AlShamaileh
Department of Chemistry, The University of Jordan, Amman, Ajman, Jordan

Othmane Amadine
VARENA Center, MAScIR Foundation, Rabat Design, Rabat, Morocco

Subin Balachandran
School of Biosciences, Mahatma Gandhi University, Kottayam, Kerala, India

Vinaya Chandran
School of Biosciences, Mahatma Gandhi University, Kottayam, Kerala, India

Karim Danoun
VARENA Center, MAScIR Foundation, Rabat Design, Rabat, Morocco

Youness Essamlali
VARENA Center, MAScIR Foundation, Rabat Design, Rabat, Morocco

Ikram Ganetri
VARENA Center, MAScIR Foundation, Rabat Design, Rabat, Morocco

Jesiya Susan George
International and Inter University Centre for Nanoscience and Nanotechnology, Mahatma Gandhi University, Kottayam, Kerala, India

Abdul Ghaffar
Department of Physics, University of Agriculture, Faisalabad, Punjab, Pakistan

Rakhimol K.R.
International and Inter University Centre for Nanoscience and Nanotechnology, Mahatma Gandhi University, Kottayam, Kerala, India

Jayachandran K.
School of Biosciences, Mahatma Gandhi University, Kottayam, Kerala, India

Nandakumar Kalarikkal
International and Inter University Centre for Nanoscience and Nanotechnology, Mahatma Gandhi University, Kottayam, Kerala, India

Jyothis Mathew
School of Biosciences, Mahatma Gandhi University, Kottayam, Kerala, India

Linu Mathew
School of Biosciences, Mahatma Gandhi University, Kottayam, Kerala, India

Stalin Nadarajan
Institute of Plant Science, ARO Volcani Center, Rishon Lezion, Israel

Muhammad Yasin Naz
Department of Physics, University of Agriculture, Faisalabad, Punjab, Pakistan

Chandra Wahyu Purnomo
Chemical Engineering Department, Universitas Gadjah Mada, Sleman, Yogyakarta, Indonesia

Maya Rajan
School of Biosciences, Mahatma Gandhi University, Kottayam, Kerala, India

Hens Saputra
Agency for the Assessment and Application of Technology, Jakarta, Indonesia

S. Shahena
School of Biosciences, Mahatma Gandhi University, Kottayam, Kerala, India

Hitha Shaji
School of Biosciences, Mahatma Gandhi University, Kottayam, Kerala, India

Shazia Shukrullah
Department of Physics, University of Agriculture, Faisalabad, Punjab, Pakistan

Reshma Soman
School of Biosciences, Mahatma Gandhi University, Kottayam, Kerala, India

Surya Sukumaran
School of Pure and Applied Physics, Mahatma Gandhi University, Kottayam, Kerala, India

Sabu Thomas
International and Inter University Centre for Nanoscience and Nanotechnology, Mahatma Gandhi University, Kottayam, Kerala, India

Remya V.R.
International and Inter University Centre for Nanoscience and Nanotechnology, Mahatma Gandhi University, Kottayam, Kerala, India

Mohamed Zahouily
VARENA Center, MAScIR Foundation, Rabat Design, Rabat, Morocco; Laboratoire de Matériaux, Catalyse et Valorisation des Ressources Naturelles, Université Hassan II-Casablanca, Morocco

CHAPTER 1

Conventional methods of fertilizer release

S. Shahena, Maya Rajan, Vinaya Chandran, Linu Mathew
School of Biosciences, Mahatma Gandhi University, Kottayam, Kerala, India

1. Introduction

The fertilizer is any kind of material that is applied to soil or to plant tissues to supply one or more plant nutrients essential to the growth of plants. It may be natural or synthetically produced [1].

Management of soil fertility has been a great problem for the farmers for thousands of years. Records showed that Egyptians, Romans, Babylonians, and early Germans were using minerals and manure to enhance the productivity of their farms. The modern science of plant nutrition started in the 19th century with the work of German chemist Justus von Liebig [1].

The Haber process and the Ostwald process developed in the 1910 and 1920s made a great revolution in the fertilizer manufacturing Industry. Ammonia (NH_3) is produced from methane (CH_4) gas and molecular nitrogen (N_2) through the Haber process, which is then converted into nitric acid (HNO_3) in the Ostwald process [2].

The nitrogen-based fertilizer production was started with Birkeland—Eyde process, which was one of the most competing industrial processes in the nitrogen-based fertilizer production. In this process, the atmospheric nitrogen (N_2) is fixed into nitric acid (HNO_3) through nitrogen fixation. The nitric acid was then used as a source of nitrate (NO_3^-) [3].

Nowadays, it has been estimated that almost half the people on the Earth are currently fed as a result of synthetic nitrogen fertilizer. The development of synthetic fertilizer has significantly supported global population growth [4]. In the last 50 years, the use of commercial fertilizers has been increasing steadily; reaching almost 100 million tons of nitrogen per year, and it is estimated that about one-third of the food produced now could not be produced without the addition of fertilizers [5]. The use of phosphate fertilizers has also increased from nine million tons per year in 1960 to 40 million tons per year in 2000. Yara International is the world's largest producer of nitrogen-based fertilizers [6,7].

Controlled Release Fertilizers for Sustainable Agriculture
ISBN 978-0-12-819555-0
https://doi.org/10.1016/B978-0-12-819555-0.00001-7

The supply of nutrients must be optimum for the maximum yield of any crop. The nutrient deficiency will result in stunting of plants and that will gradually reduce the yield by slowing down the progress of the growth cycle, causing late fruiting and delayed maturity. The ability of the crop to absorb the nutrients from the soil depends upon the biological activity. Normally the nutrient deficiency takes place during the growing season and depends on the temperature and moisture content of the soil [8]. Fertilizers enhance plant growth traditionally; either as, being additives that provide nutrients or by enhancing the effectiveness of the soil by modifying its water retention and aeration [9]. Also, an important component of the weed management program is the efficient and appropriate management of fertilizer in terms of evaluation of best source of nutrients, optimum rates of fertilization, proper timing, and suitable fertilizer placement [3,10].

The essential prerequisite for optimizing nutrient application is the detailed knowledge about the addition of nutrients, i e., the absorption of nutrients by the plants. The applied nutrients should satisfy the plant requirements and the method used for the application should minimize the leaching to the environment and thereby control the rate of environmental pollution [11]. The type of fertilizer, timing of fertilizer application, and seasonal trends are the major factors that affect the efficiency of the applied N (nitrogen) to satisfy the N demand of the crops [12,13]. The efficiency of the crops to absorb the N is influenced by the soil type, crop sequence, and the residual and mineralized N [14]. The reduction of nitrogen loss and increase in the N use efficiency can be improved by numerous strategies. For example, the use of N sources, consumption of slow-release fertilizer, proper placement techniques, and also by the use of N inhibitors [15−17].

The plant metabolism is coupled with the availability of the N sources because it has a fundamental role in the plant metabolism. It is necessary to optimize the management of N resources to the cropping system to increase its N use efficiency and thereby improve the productivity [18]. Normally, this can be achieved either by increasing the production of N in the soil or by increasing the accumulation of N compounds in the edible part of the crop [15].

The nutrients required for healthy plant life are classified on the basis of the elements; but these elements are not used directly as fertilizers. The compounds containing these elements are the basis of fertilizers. The macronutrients are consumed in larger quantities by the plants. They are present in plant tissue in quantities from 0.15% to 6.0% on a dry matter (DM) (0% moisture) basis. Plants are made up of four main elements such as hydrogen, oxygen, carbon, and nitrogen. Hydrogen, oxygen, and carbon will be available in the form of water and carbon dioxide. The nitrogen is found in the atmosphere as atmospheric nitrogen which is unavailable to plants. So the nitrogen is considered as the most important fertilizer since nitrogen is present in proteins, DNA, and other components such as chlorophyll. Some bacteria and their host leguminous plants can fix atmospheric nitrogen (N_2) by converting it to ammonia. Phosphate is required for the production of DNA and ATP, the main energy carrier in cells, as well as certain lipids.

The fertilizer contains:

- Three main macronutrients:
  - Nitrogen (N): leaf growth
  - Phosphorus (P): Development of roots, flowers, seeds, fruit
  - Potassium (K): Strong stem growth, movement of water in plants, promotion of flowering and fruiting
- Three secondary macronutrients: calcium (Ca), magnesium (Mg), and sulfur (S)
- Micronutrients: copper (Cu), iron (Fe), manganese (Mn), molybdenum (Mo), zinc (Zn), boron (B). Of occasional significance are silicon (Si), cobalt (Co), and vanadium (V).

Micronutrients are required in smaller quantities, in parts-per-million (ppm) and are present at the active sites of enzymes in the plant tissues that carry out the plant's metabolism [19].

2. Classification

Fertilizers are classified in several ways.

2.1 Based on the nutrient supply

On the basis of the nutrient supply, the fertilizers can be classified into straight fertilizers and complex fertilizers.

2.1.1 Single nutrient or straight fertilizers

As the name indicates, the single nutrient or straight nutrient fertilizers provide a single nutrient to the plants (e.g., K, P, or N). Ammonia or its solutions are the widely used nitrogen-based straight fertilizers. Ammonium nitrate (NH_4NO_3) and urea are popular sources of nitrogen. Urea is having the advantage that it is solid and nonexplosive, unlike ammonia and ammonium nitrate.

The superphosphates are the main straight phosphate fertilizers. Single superphosphate (SSP) consists of 14%−18% P_2O_5, again in the forms of

$Ca(H_2PO_4)_2$ and also phosphogypsum ($CaSO4 \cdot 2H_2O$). The main constituents of triple superphosphate (TSP) are 44%–48% of P_2O_5 and no gypsum. A mixture of single superphosphate and triple superphosphate is called double superphosphate. Most of the (more than 90%) typical superphosphate fertilizer is water-soluble [20].

Muriate of Potash (MOP) is the main potassium-based straight fertilizer. Muriate of Potash consists of 95%–99% KCl, and is typically available as 0-0-60 or 0-0-62 fertilizer [21].

2.1.2 Multinutrient or complex fertilizers

The multinutrient or complex fertilizers provide two or more nutrients (e.g., N and P). The commonly used fertilizers are the complex fertilizers. Since they consist of two or more nutrient components, they are again classified into binary fertilizers and three-component fertilizers or NPK fertilizers [20].

2.1.2.1 Binary (NP, NK, and PK) fertilizers

Since they provide both nitrogen and phosphorus to the plants they are called NP fertilizers, e.g., monoammonium phosphate (MAP) and diammonium phosphate (DAP). The active ingredient in MAP is $NH_4H_2PO_4$ and the active ingredient in DAP is $(NH_4)_2HPO_4$. About 85% of MAP and DAP fertilizers are soluble in water.

2.1.2.2 NPK fertilizers

Nitrogen, phosphorus, and potassium containing fertilizers are called three-component fertilizers or NPK fertilizers.

2.2 Based on the presence or absence of carbon

2.2.1 Organic fertilizer

Organic fertilizers are recycled plant- or animal-derived matter.

2.2.2 Inorganic fertilizer

Inorganic fertilizers or synthetic fertilizers are synthesized by various chemical treatments [20].

3. Mode of application

The application rates of fertilizer depend on the soil fertility. The fertility of a soil is usually as measured by a soil test according to the particular crop. The method of applying fertilizers depends on the nature of crop plants, their nutrient needs, and the soil (Table 1.1).

Table 1.1 A comparison of different methods of fertilizer application.

Type of fertilizer	Mode of application	Advantages	Disadvantages
Solid fertilizers	**A.** Broad casting 　**a.** Basal application 　**b.** Top dressing	1. Uniform distribution. 2. Completely mix with soil. 3. Supplying nitrogen in readily available form to growing plants.	1. The plants in the field cannot fully utilize the fertilizers. 2. Due to the presence of fertilizer all over the field, the weeds also absorb the nutrients and the weed growth is also stimulated by the fertilizer. 3. Large amount of the fertilizer is needed and nutrients are fixed in the soil.
	B. Placement 　**a.** Plough sole placement 　**b.** Deep placement 　**c.** Localized placement 　　• Drilling 　　• Side dressing	1. Suitable for dry land areas and paddy fields. 2. Used for the placement of ammoniacal nitrogenous fertilizers particularly in the root zone soil. 3. Prevents the loss of nutrients by runoff. 4. Only adequate amount of fertilizer is applied to the soil close to the seed or to the roots of growing plants. 5. Suitable for the application of phosphatic and potassic fertilizers in the case of cereal crops. 6. There is minimum contact between the soil and the fertilizer.	1. Due to the higher concentration of the soluble salts, the germinated seeds and young plants may get damaged.

Continued

Table 1.1 A comparison of different methods of fertilizer application.—cont'd

Type of fertilizer	Mode of application	Advantages	Disadvantages
		7. The nutrients are available only for the crop plants and the weeds all over the field cannot make use of the fertilizers. 8. Higher residual response of fertilizers.	
	C. Band placement **a.** Hill placement **b.** Row placement	1. Application of fertilizers in orchards. 2. Nutrients easily available to the crop plants. 3. Labor saving.	1. The seasonal change the mode of application of fertilizer.
	D. Pellet application	1. Used for the placement of nitrogenous fertilizers in the paddy fields.	1. Significant reduction of fertilizers in the flood water.
Liquid fertilizers	**A.** Starter solutions	1. The application of solution to young vegetable plantlets particularly at the time of transplantation. 2. Helps in rapid establishment and quick growth of the seedlings.	1. Additional labor needed. 2. Higher fixation rate of phosphate. 3. The plants get "shocked," due to the damage to or breaking of roots.
	B. Foliar application	1. The leaves can easily and directly absorb the nutrients through their stomatal openings and also through the epidermis. 2. An effective method of fertilization. 3. Leaves can easily absorb several nutrient elements. 4. The concentration of the fertilizer solution can be controlled manually.	

	5. Reduce the damaging and burning of the leaves.
	6. The minor nutrients such as iron, copper, boron, zinc, and manganese can be easily applied by foliar application.
	7. The insecticides are also applied along with fertilizers by foliar application.
C. Injection into soil	1. The loss of plant nutrients can be prevented.
	2. Used for controlling or eradicating the weeds.
	3. Very effective and safer for the underground water in terms of contamination and causes no or less health hazards.
D. Aerial application	1. Useful for hilly areas, forest lands, grass lands, sugarcane fields, etc.
	2. The loss of fertilizer is considerably low in this method.
E. Fertigation	1. Less nutrient loss and is well controlled.
	2. Both the water-soluble solid and liquid fertilizers can be applied along with irrigation water.

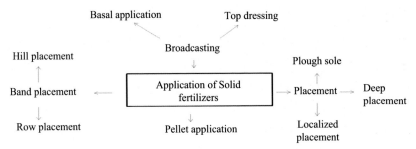

Figure 1.1 Methods of solid fertilizer application.

Fertilizers are applied to crops both in the form of solids and liquids. Most of the fertilizers are applied in the form of solids (e.g., urea, diammonium phosphate, and potassium chloride). Solid fertilizer is typically used in granulated or powdered form. It is also available in the form of prills or solid globules [22]. Liquid fertilizers comprise anhydrous ammonia, aqueous solutions of ammonia, and aqueous solutions of ammonium nitrate or urea. The concentrated liquid fertilizers can be diluted with water (e.g., UAN). Its more rapid effects and easier coverage are the advantages of liquid fertilizer [9](Fig. 1.1).

3.1 Application of solid fertilizers

3.1.1 Broadcasting

The spreading of fertilizer all over the field in a uniform manner is known as broadcasting. A separate operation in addition to seeding is required in the broadcasting mode of fertilizer application. The fertilizer may be spread on the surface of the soil itself, with or without incorporation into the soil, or it may be placed below the soil surface in closely spaced rows by the use of a fertilizer drill [8].

Normally the fertilizer used for this kind of application is in an insoluble form; especially, insoluble phosphatic fertilizer such as rock phosphate is used for broadcasting mode of application. This method is suitable for crops with dense stand. The plant roots will permeate the whole volume of the soil. Large doses of fertilizers are needed for the application.

There are two methods of broadcasting method of application, namely broadcasting at sowing or planting (basal application) and top dressing.

3.1.1.1 Broadcasting at sowing or planting (basal application)

The main objective of the basal application is the application of fertilizers at sowing time for a uniform distribution. Thus the fertilizer will be spread over the entire field and completely mix with soil.

Boron fertilizers are generally applied by broadcast method. Normally they are incorporated prior to seeding for crops not planted in rows. Boron is applied by broadcast method in plants such as legumes and grasses and broadcast methods are more effective in trees and grape vines and also in the cases of coarser-textured soils [23].

Blackshaw et al. [24] reported that N uptake by green foxtail throughout the growing season was often greater from surface broadcast than from surface pools or point-injected N. To keep abreast of increasing population, the rice production of Asia must be increased up to 2.2%–2.8% annually. The efficiency of fertilizer N can be increased by the improvement of timing and application methods, particularly through the better incorporation of basal fertilizer N without standing water [25].

3.1.1.2 Top dressing

The nitrogenous fertilizers are normally applied closely in crops like paddy and wheat, with the objective of supplying nitrogen in readily available form to growing plants. This kind of application of nitrogen fertilizer is known as top dressing.

To improve rice yield and the nitrogen availability to the plants, top dressing is recommended to the lower soil layer for Japonica rice [26–28], new high yielding rice varieties such as Indica type [29], and large grain type varieties [28,30].

In the case of rice varieties, the timing of top dressing with high nitrogen (HN) to produce high rice yield is not fully understood. Matsushima [26] observed that topdressing at 30 days before heading (30 DBH) resulted in worse plant type and yield reduction due to the elongation of the lower internodes and upper leaves. In a model experiment with Japonica type variety Koshihikari, Matsuba [31] indicated that the topdressing at 30 DBH did not elongate the lower internodes. In the case of Takanari, an Indica variety, the topdressing at 30 DBH did not worsen the plant type but did increase the sink size in high-yielding varieties [26]. Fukushima et al. [28] suggested that the new type rice variety Bekoaoba will increase its sink size and the rice yield by top dressing at 30 DBH or early top dressing leading to short culms and erect leaf.

In Bangladesh, crystal urea is normally applied as top dressing. It decreases yield by misbalancing the yield components. Usually this problem is prevented by the application of super granules of urea (USG); since the USG have the ability to minimize the loss of N from soil, thus effectively increasing up to 20%–25% [32].

Disadvantages of broadcasting: The plants in the field cannot fully utilize the fertilizers as they move laterally over long distances. Due to the presence of fertilizer all over the field, the weeds also absorb the nutrients and the weed growth is also stimulated by the fertilizer. Large amount of the fertilizer is needed and nutrients are fixed in the soil. They may come in contact with a large mass of soil.

3.1.2 Placement
The placement of the fertilizer in soil at a specific place with or without reference to the position of the seed is referred as placement method of fertilizer application. The placement method is normally recommended in conditions where the quantity of the fertilizer is small and the soil has low fertility. It can be applied also in case of plants with poorly developed roots. The phosphatic and potassic fertilizers are normally applied by placement method.

The commonly used methods of placement methods are plow sole placement, deep placement, and localized placement.

3.1.2.1 Plough sole placement
During the process of ploughing, the fertilizer is placed at the bottom of the plough furrow in the form of a continuous band so that every band is covered as the next furrow is turned. This method is suitable for dry land where the surface soils becomes quite dry up to a few centimetres from the bottom soil and have a heavy clay pan just below the plough sole layer.

3.1.2.2 Deep placement
It is used for the placement of ammoniacal nitrogenous fertilizers in the reduction zone of the soil, particularly in the root zone. This method is especially suitable for paddy fields. The main advantage of this method is to prevent the loss of nutrients by run-off.

The Japanese used different methods of N fertilizer application to minimize the loss of N through volatization, denitrification, leaching, etc. Based on the Japanese concept of deep point placement of fertilizer N in transplanted rice, IFDC (International Fertilizer Development Center) implemented the use of super granules of urea (USG) to achieve the same agronomic benefits in 1975 [33].

The loss of nitrogen is greater in rice (*Oryza sativa* L.) fields especially in the irrigated rice cropping systems with very poor water control. Bandaogo et al. [34] conducted field experiments on Sourou Valley in Burkina Faso of

West Africa on the wet season of 2012 and dry season 2013. The studies were based on the effect of fertilizer N in the form of prilled urea which is applied by broadcasting and briquettes in the form of USG, which is applied by FDP (fertilizer deep point placement). The results indicated that FDP is genotype and season specific and it can be suggested as an alternative method for the farmers to improve nitrogen use efficiency (NUE) in the irrigated rice cropping system.

Nitrogen is found to be an essential nutrient for the growth of the rice plants. Usually prilled urea (PU) was applied by broadcasting because it is considered as a fast releasing source of nitrogen. But in flooded rice fields, it can be lost by ammonia volatilization, immobilization, denitrification, and surface runoff [35]. To overcome this problem, the USG can be applied by deep placement. USG is a slow-releasing nitrogenous fertilizer and it reduces the N loss and also improves the N use efficiency of wetland rice [36].

Craswell and De Datta [37] recommended that the application of urea super granules up to 10 cm depth by deep placement method minimizes the loss of N. The broadcasting of the same at the soil surface level causes losses up to 50%. The urea supergranules can be also placed in the root zone. This kind of application also increases the rice yield and nitrogen use efficiency of the plants [36,38].

Roger et al. [39] studied the effect of various kinds of nitrogen fertilizer application on the algal flora and also about the biological nitrogen fixation (acetylene-reducing activity) in a wetland rice soil by both pot and field analysis. According to them, the deep placement of USG is found to be better than the broadcast application of urea, since the broadcast application of urea inhibited the nitrogen fixation by favoring the growth of the green algae. However, the deep placement of USG does not report any negative effect on the nitrogen fixing blue-green algae. On the other hand, it permits acetylene reducing activity on the soil surface and it does not inhibit the growth of nitrogen fixing bacteria.

3.1.2.3 Localized placement
In order to supply the nutrients, adequate amount of fertilizer is applied to the soil close to the seed or to the roots of growing plants. The common methods used for the placement of fertilizers or nutrients are:

3.1.2.3.1 Drilling In this method a seed-cum- fertilizer mode of drilling is used for the application of fertilizer during the sowing time itself. The fertilizer and the seed are placed in the same row but the depth is different. The method is suitable in the case of cereal crops, especially for the application of phosphatic and potassic fertilizers. Due to the higher concentration of the soluble salts, the germinated seeds and young plants may get damaged. This is the greatest disadvantage of this method.

3.1.2.3.2 Side dressing The fertilizers are spread between the rows and around plants. In the case of crops like maize, sugarcane, cotton, etc., the fertilizer is applied by hand in between the rows. But the fertilizers are placed around the trees like mango, apple, grapes, papaya, etc. Adiaha and Agba [40] found out that among the four methods used (broadcasting, ring application, hole application, and liquid application) for the cultivation of maize plants (*Zea mays* L.), ring method seems to be appropriate for maize production at 1 m spacing between plants on bed.

Advantages of placement of fertilizers:

The main advantages of placement mode of fertilizer applications are as follows:

(1) There is minimum contact between the soil and the fertilizer.

(2) The fixation of nutrient is greatly reduced.

(3) The nutrients are available only for the crop plants and the weeds all over the field cannot make use of the fertilizers.

(4) Higher residual response of fertilizers.

(5) Higher utilization of fertilizers by the plants.

(6) Loss of fertilizer, for example, loss of nitrogen, by leaching is reduced.

(7) Immobile phosphates are better utilized when placed.

3.1.3 Band placement

The band placement refers to the fertilizer placement in the form of bands. In the case of band placement, the fertilizers can be applied by hill placement method or by row placement method.

3.1.3.1 Hill placement

The fertilizer is applied close to the plants on one or both sides of the plants as bands but the length and the breadth of the band is varied with the nature of the crop. This method is common for the application of fertilizers in orchards.

Ibrahim et al. [41] reported that the hill placement of manure and fertilizers like DAP (diammonium phosphate) and NPK improved the yield and efficiency of millets in Sahelian agro-ecological area of Niger.

3.1.3.2 Row placement

The fertilizer is applied in continuous bands on one or both sides of the row in which the plants are planted. This method is common in the case of crops like sugarcane, potato, maize, cereals, etc.

There will be a fertilizer attachment on the planter in cases of hill or row placement. So there is no need for a separate operation. These methods are labor saving and in some rare cases, the seeds and the fertilizer will be incorporated in soil separately by using machine or by hand [8]. Under drier conditions, small grains of fertilizers have better response to band applications [42].

The maximum uptake of applied nutrients depends on the zone of nutrient concentration and root activity [43]. Chaudhary and Prihar [44] observed that if the fertilizer was placed 20 cm below the seeds, the nutrient uptake was found to be increased and the seedling growth was also found to be faster and resulted in higher grain yield in the case of maize (*Zee mays* L.) and wheat (*Triticum aestivum* L.). This result indicates that the band placement is more effective than the broadcast mode of fertilizer application. If the growths of the roots are found only in the surface layers, it is advised that the fertilizer should not be broadcasted to avoid surface drying.

The observations in *Oryza sativa* L. made by Cao et al. [45] revealed that the seasonal change like the drifting of dry season into wet season may also affect the mode of application of fertilizer. The result showed that the uniform placement and the point placement of USG gave the highest yield (6.4 Mg/Ha) with efficiency of 51 kg rough rice/kg N in the dry season. At the same time, in wet season, by point placement with USG produced highest yield of 4.4 Mg/Ha with efficiency of 40 kg rough rice/kg N. The records of the rice crop at harvest, both from grain and straw suggested that the fertilizer N uptake was high as 75% in dry season and 65% as in wet season.

3.1.4 Pellet application

It is mainly used for the placement of nitrogenous fertilizers in the paddy fields. The fertilizers are applied in the form of pellets of about 2.5−5 cm deep between the crops. The small pellets of convenient sizes of fertilizers are made and mixed with the soil in the ratio of 1:10. The pellets are deposited in the mud of paddy fields.

Schnier et al. [46] conducted a study in transplanted rice and direct seed-flooded rice to evaluate the effect of time and method of fertilizer N application on grain yield and N-use efficiency by using ^{15}N labeled urea.

The conventional application of methods including broadcasting and incorporation were compared with band placement of liquid urea and point placement of USGs. The study indicated that the band or point placement results in significantly greater grain yield with significant reduction in the partial pressure of $NH_3(pNH_3)$ in flood water.

Increased use of N fertilizers in agricultural practices mostly results in environmental contamination by leaching N into plant-soil system. Reza–Bhageri et al. [47] conducted a field experiment to study the effect of pellet fertilizer application in corn fields. According to them, the use of pellet form of fertilizer is an alternative and better method for the application of N to the corn field which results in maximum grain yield and highest grain protein content. The N application in the form of pellet releases its content slowly and due to continuous nutrient release, the plant can uptake N at different stages of its growth.

3.2 Application of liquid fertilizers and water-soluble granular fertilizers

Liquid fertilizers and water-soluble granular fertilizers can be applied by the following methods:

3.2.1 Starter solutions

A solution of N, P_2O_5, and K_2O in the ratio of 1:2:1 and 1:1:2 applied to young vegetable plantlets particularly at the time of transplantation is normally referred to as starter solution. This method helps in rapid establishment and quick growth of the seedlings. The additional labor and higher fixation rate of phosphates are the two major disadvantages of starter solution method of fertilizer application.

During the time of transplantation of plants, the plants get "shocked" due to the damaged or broken roots. As a result, the uptake of water and nutrients by the roots will be restricted and stunted growth or death of the plants may occur finally. Replacing the use of pure water with dilute solutions containing plant nutrients often reduces the shock of transplanted plants resulting in faster establishment of plants [48].

To find out the best crop management practices with starter solutions, Susila et al. [49] planted chilli pepper plants (*Capsicum annuum* L.) in polyethylene mulched on inceptisol soil with pH of 5.5, low organic (1.54%), N content of 0.12% (very low), K content of 0.29 me (100g) –1, but with very high soil P_2O_5 concentration (19.2 ppm) concentration. The plants were also treated with cow manure, organic starter solutions, inorganic starter solutions, and standard inorganic fertilizer. The result indicated that the application of cow manure, standard inorganic fertilizer, or organic starter solutions reduced total unmarketable yield, enhanced the fruit quality, and improved marketable yield.

The AVRDC, 2004 (Asian vegetable Research and Development Center, Shanhua, Taiwan) reported that the supplementation of liquid NPK (nitrogen, phosphorous, and potassium) as starter solutions enhance the early growth of cherry tomatoes and cabbages grown by using organic fertilizers. They found that this is an effective technique, i e., the application of liquid fertilizer along with the organic fertilizer because it enhances N, P, K uptakes by the roots, increases plant dry weight, and also promotes the early growth of the crops with early fast growth rates. They concluded that the starter solutions may accelerate the root growth and thereby increase the efficiency of absorbing more nutrients from soil and organic fertilizers.

Gorden and Pierzynski [50] found that the use of starter solutions containing N and P consistently increased grain yields, reduced the number of thermal units required for plant emergence to maturity, decreased grain moisture content at the time of harvest, and increased total P uptake of corn.

3.2.2 Foliar application
The application of liquid fertilizers directly to the leaf surface by using spraying method is known as *foliar application*. The leaves can easily and directly absorb the nutrients through their stomatal openings and also through the epidermis. This will be an effective method of fertilization [51]. The fertilizer solutions containing one or more nutrients will be applied on the foliage of the growing plants. Since the nutrients are sprayed only after dissolving them in water, the leaves can easily absorb several nutrient elements. The concentration of the fertilizer solution can be controlled manually and this will reduce the damaging and scorching of the leaves. The minor nutrients such as iron, copper, boron, zinc, and manganese can be easily applied by foliar application. The insecticides are also applied along with fertilizers by foliar application. Foliar fertilizers can be applied directly to the leaves. The water–soluble straight nitrogen fertilizers are also used for foliar application. It is especially used for high value crops such as fruits [9].

To facilitate the optimal N management, urea can be applied by foliar application. This mode of application increases yield and minimizes the loss of N to the environment [52]. The plants have the ability to absorb foliar applied urea rapidly and hydrolyze the urea in cytosol [53,54]. Thus the foliar applied urea as N source is a common practice [55,56]. The season [57,58] and the physiological effects caused by the foliar applied urea as N source may vary with the cultivar [59].

The foliar applied urea induces a positive effect in the wheat cultivation by increasing the photosynthetic rate and urease enzyme activities. But in the case of soybeans, the yields were inconsistent depending on the year and cultivar used by the foliar application [59].

Nicoulaud and Bloom [53] observed that foliar applied urea promoted the growth of T-5 tomatoes (*Lycopersicon esculentum* Mill. cv. T5) because it absorbed 75% of the urea applied within 12 h and 99% within 24 h of application. The results concluded that (a) the foliar applied urea can supply the required N to sustain the growth of the seedlings, (b) to alleviate the N deprivation, the applied urea is absorbed by the seedlings as fast as possible, (c) the failure in the promotion of rapid growth by urea is probably due to phytotoxicity.

Heumann et al. [60] prepared a sustained release fertilizer composition. In this, the water-soluble granular fertilizers are subjected first to accretive granulation under accretive granulation conditions with a melt solution or slurry of one or several fertilizers. A second stage coating supplied these treated grains with a suitable almost impermeable envelopment. In this process the granules are subjected to various conditions of temperature and accretion under which they can be regranulated. The conditions will be depending upon the chemical composition and the size of the water-soluble granules being treated and the intended method of accretive granulation. One or more preponderantly water-soluble fertilizer composition such as urea, ammonium-, potassium-, NP-, NK-, PK-, or NPK-fertilizers can be used in the granular preparation. The fertilizers that are treated include those prepared from conventional methods such as crystallization, pelletizing, graining, crushing, pressing, or spray drying. Particularly contemplated compositions for treatment include fertilizer prills [61].

The zinc deficiency of the soil can be corrected by foliar spraying and it will result in the improvement in the growth of the plant, yield, and seed quality [62,63]. Thalooth et al. [64] reported that foliar spraying with zinc resulted in the improved yield of the sunflower plants. Gobarah et al. [65] concluded that the foliar application of 1.00 g/L zinc and P_2O_5/fad (phosphorous pentoxide) enhanced the growth, yield, and seed quality of the groundnuts.

The lime-induced iron deficiency chlorosis in citrus plants can be treated by foliar spraying of iron compounds. The photosynthesis and peroxidase activity can also be restored by the treatment [66].

3.2.3 Injection into soil/plants

The liquid fertilizers can be injected into the soil by either pressure or nonpressure types. Nonpressure solutions may be applied on the surface or in furrows. The loss of plant nutrients can be prevented by the injection of liquid fertilizers into the soil. For example, anhydrous ammonia placed in narrow furrows at a depth of 12−15 cm will be covered suddenly to prevent loss of ammonia.

The plant roots take up only a very small portion of soil-added fertilizers. In most of the cases, high soil permeability allows the loss of nutrients especially by fast leaching of the fertilizers to the underground water [67]. The fertilizers are also lost by volatization, especially N [68]. The addition of nutrients like phosphorous and micronutrients in the form of dissolved compounds also prevents the absorption of added fertilizers by the roots [66].

The fertilizers also can be directly applied to the tree trunk. The main advantage of the injection of fertilizers directly in to the plant trunk is the treatments used for controlling or eradicating the weeds can be avoided because the weeds cannot compete with the crop plants for the available nutrients. Shaaban [10] recommended the application of injection fertilization, especially directly to trunks of mango and grapevine. It is a very effective method for the nutrient supply as well as found to be safer for the underground water in terms of contamination and causes no or less health hazards.

3.2.4 Aerial application

The application of liquid fertilizer using aircraft in areas where ground application is not possible is known as *aerial application*; for example, in hilly areas, forest lands, grasslands, sugarcane fields, etc. The loss of fertilizer is considerably low in this method.

The aerial application of superphosphate in *Pinus radiata* of Forest Research Institute in New Zealand recommended that the application of fertilizers by airplane is an accepted tool of management of *P. radiata* on phosphate deficient soils and have the benefits of optimum time, rate, and frequency of application [69].

3.2.5 Fertigation

There are two types of irrigation practices used for the application of fertilizers along with irrigation, i e., pressurized irrigation and drip irrigation. The major disadvantage of normal irrigation practice is the efficiency of water use in this mode is low. But pressurized irrigation practices normally have higher water use efficiency, less nutrient loss, and are well controlled. The initial cost and maintenance costs and expertise are the major constraints of the pressurized irrigation practices. The most appropriate method of water and nutrient application is the drip irrigation or fertigation [70].

Both the water-soluble solid or liquid fertilizers can be applied along with irrigation water. The combined application of water-soluble solid or liquid fertilizers with irrigation water through pressurized irrigation system is known as fertigation. The nitrogenous fertilizers such as urea and other ammoniac fertilizers which are easily soluble in water are applied along with irrigation water. The application of fertigation will increase yield and minimize soil and water pollution. The loss of fertilizer is considerably low in this method [71].

The nutrient addition by fertigation is determined by the concentration of the nutrients in irrigation water, the nutrient absorption by the plant, the rate of evapotranspiration, and the reaction (precipitation or fixation) by the growth medium [11].

Hebbar et al. [72] studied the effect of fertigation and evaluated the sources and level of fertilizer application on growth, yield, and fertilizer use efficiency of hybrid tomato in red sandy loam soil. The data showed a significantly higher production of total dry matter (TDM) and leaf area index (LAI). The chlorophyll concentration was significantly higher in fertigation treatments and also resulted in lesser leaching of NO3-N and K. Root growth and uptake of NPK was also increased by WSF (water-soluble fertilizer) fertigation. The commonly used fertigation systems are pressure differential, the Venturi (vacuum), and injection pump.

In California, in the late 1960s, about 5% of the nitrogen fertilizers were applied with irrigation water [73]. A chemigation survey in USA conducted by Threadgill [74] showed that only 3.5% of them used nitrogen fertilizers along with irrigation water. Sixty-one percent used micro-irrigation systems and 43% used sprinkler systems.

4. The controlled-release fertilizers

By the conventional methods of fertilizer application, the nitrogen fertilizer is not fully absorbed by the crop plants. It will migrate to the water bodies leading to nutrient enrichment and cause negative impacts on environmental quality, ecosystem services, and biodiversity [75–77]. There has been much research conducted on technologies to improve fertilizer application practices and eliminate adverse impacts on water ecosystems [78–83].

Controlled-release fertilizers are encapsulated in a shell that degrades at a specified rate. The commonly used encapsulation material is sulfur. Thermoplastics and sometimes ethylene-vinyl acetate and surfactants can be used as a coating material to produce diffusion-controlled release of urea or other fertilizers. The use of reactive layer coating can produce thinner and cheaper membrane coatings by applying reactive monomers simultaneously to the soluble particles. The practice of application of layers of low-cost fatty acid salts with a paraffin topcoat is called *Multicote* [7].

5. Slow- and controlled-release fertilizers

Most slow-release fertilizers are derivatives of urea. This is a kind of straight fertilizer providing nitrogen, e.g., Isobutylidenediurea (IBDU) and urea-formaldehyde. They slowly convert N in the soil to free urea, which is rapidly taken up by plants. IBDU is a single compound with the formula $(CH_3)_2CHCH(NHC(O)NH_2)_2$ whereas the urea-formaldehydes consist of mixtures of the approximate formula $(HOCH_2NHC(O)NH)_nCH_2$. They are more efficient in the utilization of the applied nutrients; these technologies reduce the impact on the environment and the contamination of the subsurface water. Slow-release fertilizers are applied in various forms including fertilizer spikes, tabs, etc. The use of slow-release fertilizers reduces the problem of "burning" the plants due to excess nitrogen [7].

The slow- and controlled-release fertilizers have many disadvantages. Due to the antagonistic property of the slow-and controlled-release fertilizers, their practical applications involve only 0.15% (562,000 tons) of the fertilizer market. Even though they provide nutrition to plants, excess fertilizers can be poisonous to some plants and may result in the degradation or loss of the fertilizer. Also, microbes degrade many fertilizers by immobilization or oxidation. They can also be lost by evaporation or leaching [7].

6. Chemicals that affect nitrogen uptake

The efficiency of nitrogen-based fertilizers can be enhanced by the addition of various chemicals, and thereby the farmers can limit the polluting effects of nitrogen runoff. Nitrification inhibitors or nitrogen stabilizers suppress the conversion of ammonia into nitrate, an anion that is more prone to leaching. 1-Carbamoyl-3-methylpyrazole (CMP), dicyandiamide, nitrapyrin (2-chloro-6-trichloromethylpyridine) and 3,4-dimethylpyrazole phosphate (DMPP) are the popular inhibitors of nitrification [84].

The hydrolytic conversion of urea into ammonia, which is prone to evaporation as well as nitrification, is slowed down by the use of urease inhibitors. The ureases are enzymes that catalyze the conversion of urea to ammonia. Example of urease inhibitor is N-(n-butyl) thiophosphoric triamide (NBPT).

7. Overfertilization

Excess care must be taken while using the fertilizer because excess fertilizers can be detrimental to plants. Fertilizer burn will occur when excess fertilizer is applied that will result in damage or even death of the plant [61].

8. Conclusion

In earlier days, researches were focused mainly on maximizing yield. But the present researches aim at satisfying the ever growing demand for food and maintaining the quality of food by maximizing the quantity and quality of yields without neglecting the environment. The need to improve the fertilizer efficiency, particularly N fertilizer efficiency, is the major issue faced by the agricultural practices world over. One of the major causes of the environmental pollution is the excess use of the fertilizers. Fertilizer and plant nutrition research should be established to prevent the environmental pollution and at the same time meet the productivity and yield requirements. Controlled-release fertilizers, band deep placement methods, foliar application, and minimum tillage methods can be used to improve the efficiency of N fertilizers and should be improved and refined to reduce the environmental contamination [70]. Modern agricultural practices will be instrumental in devising such precise recommendations, which minimize transport of nutrients onto unwanted sites and generate high yields economically, with emphasis on the preservation of clean environment.

References

[1] H. Kiiski, H.W. Scherer, K. Mengel, G. Kluge, K. Severin, Fertilizers, 1. general, in: Ullmann's Encyclopedia of Industrial Chemistry, 2000, pp. 1—30.
[2] T.I. Williams, A Short History of Twentieth-Century Technology C. 1900-c. 1950, Oxford University Press, 1982, ISBN 978-0-19-858159-8, pp. 134—135.

[3] A.J. Ihde, The Development of Modern Chemistry, Courier Corporation, 1984, ISBN 978-0-486-64235-2, p. 678.

[4] J.W. Erisman, M.A. Sutton, J. Galloway, Z. Klimont, W. Winiwarter, How a century of ammonia synthesis changed the world, Nat. Geosci. 1 (10) (2008) 636—639.

[5] A.D. Glass, Nitrogen use efficiency of crop plants: physiological constraints upon nitrogen absorption, Crit. Rev. Plant Sci. 22 (5) (2003) 453—470.

[6] C.P. Vance, C. Uhde-Stone, D.L. Allan, Phosphorus acquisition and use: critical adaptations by plants for securing a nonrenewable resource, New Phytol. 157 (3) (2003) 423—447.

[7] P. Weerasinghae, K. Prapagar, K.M.C. Dharmasena, Nitrogen Release Patterns of Urea and Nano Urea Fertilizer Under Two Contrasting Soil Moisture Regimes, 2016.

[8] R.M. Salter, Methods of Applying Fertilizers, US Department of agriculture, 1938.

[9] H. Dittmar, M. Drach, R. Vosskamp, M.E. Trenkel, R. Gutser, G. Steffens, Fertilizers, 2.Types, in: Ullmann's Encyclopedia of Industrial Chemistry, 2009.

[10] M.M. Shaaban, Injection fertilization: a full nutritional technique for fruit trees saves 90-95% of fertilizers and maintains a clean environment, Fruit Veg. Cereal Sci. Biotechnol. 3 (1) (2009) 22—27.

[11] J. Hagin, A. Lowengart, Fertigation for minimizing environmental pollution by fertilizers, in: Fertilizers and Environment, Springer, Dordrecht, 1996a, pp. 23—25.

[12] B. Borghi, Nitrogen as determinant of wheat growth and yield, in: Wheat: Ecology and Physiology of Yield Determination, Food Products Press, New York (USA), 1999, pp. 67—84.

[13] K. Blankenau, H.W. Olfs, H. Kuhlmann, Strategies to improve the use efficiency of mineral fertilizer nitrogen applied to winter wheat, J. Agron. Crop Sci. 188 (3) (2002) 146—154.

[14] L. López-Bellido, R.J. López-Bellido, R. Redondo, Nitrogen efficiency in wheat under rainfed mediterranean conditions as affected by split nitrogen application, Field Crops Res. 94 (1) (2005) 86—97.

[15] F.G. Viets, The Plant's Need for and Use of Nitrogen, 1965, pp. 503—549. Soil nitrogen.

[16] J.H.G. Slangen, P. Kerkhoff, Nitrification inhibitors in agriculture and horticulture: a literature review, Fert. Res. 5 (1) (1984) 1—76.

[17] J.R. Freney, C.J. Smith, A.R. Mosier, Effect of a new nitrification inhibitor (wax coated calcium carbide) on transformations and recovery of fertilizer nitrogen by irrigated wheat, Fert. Res. 32 (1) (1992) 1—11.

[18] R. Novoa, R.S. Loomis, Nitrogen and plant production, Plant Soil 58 (1—3) (1981) 177—204.

[19] H.A. Mills, J.B. Jones Jr., Plant Analysis Handbook II: A Practical Sampling, Preparation, Analysis, and Interpretation Guide, 1996. No. 581.13 M657).

[20] J.B. Jones Jr., Plant Nutrition and Soil Fertility Manual, CRC press, 2012.

[21] V. Smil, Enriching the Earth: Fritz Haber, Carl Bosch, and the Transformation of World Food Production, MIT press, 2004.

[22] R.J. Haynes, R. Naidu, Influence of lime, fertilizer and manure applications on soil organic matter content and soil physical conditions: a review, Nutr. Cycl. Agroecosyst. 51 (2) (1998) 123—137.

[23] D.C. Martens, D.T. Westermann, Fertilizer Application for Correcting Micronutrient Deficiencies, 1991.

[24] R.E. Blackshaw, G. Semach, H.H. Janzen, Fertilizer application method affects nitrogen uptake in weeds and wheat, Weed Sci. 50 (5) (2002) 634—641.

[25] S.K. De Datta, Improving nitrogen fertilizer efficiency in lowland rice in tropical Asia, in: Nitrogen Economy of Flooded Rice Soils, Springer, Dordrecht, 1986, pp. 171—186.

[26] S. Matsushima, Technology for improving rice cultivation, Yokendo, Tokyo, 1973, pp. 1—393.

[27] A. Fukushima, Effects of timing of nitrogen topdressing on morphological traits in different rice varieties, Jpn. J. Crop Sci. 76 (2007) 18—27.

[28] A. Fukushima, H. Shiratsuchi, H. Yamaguchi, A. Fukuda, Effects of nitrogen application and planting density on morphological traits, dry matter production and yield of large grain type rice variety Bekoaoba and strategies for super high-yielding rice in the Tohoku region of Japan, Plant Prod. Sci. 14 (1) (2011) 56—63.

[29] Y. San-Oh, Y. Mano, T. Ookawa, T. Hirasawa, Comparison of dry matter production and associated characteristics between direct-sown and transplanted rice plants in a submerged paddy field and relationships to planting patterns, Field Crop. Res. 87 (1) (2004) 43—58.

[30] T. Mae, A. Inaba, Y. Kaneta, S. Masaki, M. Sasaki, M. Aizawa, A. Makino, A large-grain rice cultivar, Akita 63, exhibits high yields with high physiological N-use efficiency, Field Crop. Res. 97 (2—3) (2006) 227—237.

[31] K. Matsuba, A new morphogenetic model on the most suitable leaf-internode unit and developmental stage for controlling plant types in rice (*Oryza sativa*) cultivation, Jan. J. Crop Sci. (2000) (Japan).

[32] M. Hasanuzzaman, K. Nahar, M.M. Alam, M.Z. Hossain, M.R. Islam, Response of transplanted rice to different application methods of urea fertilizer, Growth 5 (2009) 6.

[33] N.K. Savant, P.J. Stangel, Deep placement of urea supergranules in transplanted rice: principles and practices, Fert. Res. 25 (1) (1990) 1—83.

[34] A. Bandaogo, F. Bidjokazo, S. Youl, E. Safo, R. Abaidoo, O. Andrews, Effect of fertilizer deep placement with urea supergranule on nitrogen use efficiency of irrigated rice in Sourou Valley (Burkina Faso), Nutr. Cycl. Agroecosyst. 102 (1) (2015) 79—89.

[35] A. Sreenivasan, V. Subrahmanyan, Biochemistry of water-logged soils. Part IV. Carbon and nitrogen transformations, J. Agric. Sci. 25 (1) (1935) 6—21.

[36] M.S. Hasan, S.M.A. Hossain, M. Salim, M.P. Anwar, A.K.M. Azad, Response of hybrid and inbred rice varieties to the application methods of urea supergranules and prilled urea, Pakistan J. Bio. Sci. 5 (7) (2002) 746—748.

[37] E.T. Craswell, S.K. De Datta, Recent Developments in Research on Nitrogen Fertilizers for rice. Soil and Fertilizer Forum of Thailand, 1980.

[38] S.K. Sharma, Present status of intermediate technology in rice production including the possible utilization of blue green algae and azolla, paper presented at the All-India Rice Workshop, held at the Directorate of Rice Research Rajendranagar, Hyderabad during 12-15 April 1985, Indian J. Asril. Sci 59 (3) (1989) 154—156.

[39] P.A. Roger, S.A. Kulasooriya, A.C. Tirol, E.T. Craswell, Deep placement: a method of nitrogen fertilizer application compatible with algal nitrogen fixation in wetland rice soils, Plant Soil 57 (1) (1980) 137—142.

[40] M.S. Adiaha, O.A. Agba, Influence of different methods of fertilizer application on the growth of maize (*Zea mays* L.). for increase production in south Nigeria, World Sci. News 54 (2016) 73—86.

[41] A. Ibrahim, R.C. Abaidoo, D. Fatondji, A. Opoku, Hill placement of manure and fertilizer micro-dosing improves yield and water use efficiency in the Sahelian low input millet-based cropping system, Field Crops Res. 180 (2015) 29—36.

[42] G.W. Randall, R.G. Hoeft, Placement methods for improved efficiency of P and K fertilizers: a review, J. Prod. Agric. 1 (1) (1988) 70—79.

[43] N.K. Fageria, V.C. Baligar, C.A. Jones, Growth Ad Mineral Nutrition of Field Crops, Marcel and Dekker, 1991.

[44] M.R. Chaudhary, S.S. Prihar, Comparison of banded and broadcast fertilizer applications in relation to compaction and irrigation in maize and wheat, Agron. J. 66 (4) (1974) 560—564.

[45] Z.H. Cao, S.K. De Datta, I.R.P. Fillery, Effect of placement methods on floodwater properties and recovery of applied nitrogen (15N-labeled urea) in wetland rice 1, Soil Sci. Soc. Am. J. 48 (1) (1984) 196—203.

[46] H.F. Schnier, M. Dingkuhn, S.K. De Datta, E.P. Marqueses, J.E. Faronilo, Nitrogen-15 balance in transplanted and direct-seeded flooded rice as affected by different methods of urea application, Biol. Fertil. Soils 10 (2) (1990) 89—96.

[47] G.A. Reza-Bagheri, H. Mohammad, A.S. Zinol, H. Mehdi-Younessi, The effect of pellet fertilizer application on corn yield and its components, Afr. J. Agric. Res. 6 (10) (2011) 2364—2371.

[48] C.H. Ma, T. Kalb, Development of starter solution technology as a balanced fertilization practice in vegetable production, in: International Symposium towards Ecologically Sound Fertilisation Strategies for Field Vegetable Production, vol. 700, June 2004, pp. 167—172.

[49] A.D. Susila, C.H. Ma, M.C. Palada, Application of starter solution increased yields of chili pepper (Capsicum annuum L.), J. Agron. Indones. (Indones. J. Agron.) 39 (1) (2011).

[50] W.B. Gordon, G.M. Pierzynski, Corn hybrid response to starter fertilizer combinations, J. Plant Nutr. 29 (7) (2006) 1287—1299.

[51] G. Kuepper, Foliar Fertilization, NCAT agriculture specialist, ATTRA Publication#, 2003. CT13.

[52] M. Giroux, Effetsd'applicationd'uree au sol et au feuillagesur le redement, le poidsspécifique et la nutrition azotée de la pomme de terre, Nat. Can. 111 (1984) 157—166.

[53] S.H. Wittwer, M.J. Bukovac, H.B. Tukey, Advances in foliar feeding of plant nutrients, in: M.H. McVickar, G.L. Bridger, L.B. Nelson (Eds.), Fertilizer Technology and Usage, Soil Science Society of America, Madison, WI, 1963, pp. 429—455.

[54] B.A. Nicoulaud, A.J. Bloom, Absorption and assimilation of foliarly applied urea in tomato, J. Am. Soc. Hortic. Sci. 121 (6) (1996) 1117—1121.

[55] F.E. Below, S.J. Crafts-Brandner, J.E. Harper, R.H. Hageman, Uptake, distribution, and remobilization of 15N-labeled urea applied to maize canopies 1, Agron. J. 77 (3) (1985) 412—415.

[56] D.C. Bowman, J.L. Paul, The foliar absorption of urea-N by tall fescue and creeping bentgrass turf, J. Plant Nutr. 13 (9) (1990) 1095—1113.

[57] Z. Han, X. Zeng, F. Wang, Effects of autumn foliar application of 15n-urea on nitrogen storage and reuse in apple, J. Plant Nutr. 12 (6) (1989) 675—685.

[58] D. Ippersiel, I. Alli, A.F. MacKenzie, G.R. Mehuys, Nitrogen distribution, yield, and quality of silage corn after foliar nitrogen fertilization, Agron. J. 81 (5) (1989) 783—786.

[59] J. Peltonen, Interaction of late season foliar spray of urea and fungicide mixture in wheat production, in: Plant Production on the Threshold of a New Century, Springer, Dordrecht, 1994, pp. 397—399.

[60] H. Heumann, H. Hahn, W. Hilt, H. Liebing, M. Schweppe, U.S. Patent No. 4142885, U.S. Patent and Trademark Office, Washington, DC, 1979.

[61] H. Garrett, Organic Lawn Care: Growing Grass the Natural Way, University of Texas Press, 2014.

[62] D.S. Darwish, E.G. El-Gharreib, M.A. El-Hawary, O.A. Rafft, Effect of some macro and micronutrients application on peanut production in a saline soil in El-Faiyum Governorate, Egypt. J. Appl. Sci. 17 (2002) 17—32.

[63] A.A. Ali, S.A.E. Mowafy, Effect of different levels of potassium and phosphorus fertilizers with the foliar application of zinc and boron on peanut in sandy soils, Zagazig, Egypt. J. Agric. Res. 30 (2) (2003) 335—358.

[64] A.T. Thalooth, N.M. Badr, M.H. Mohamed, Effect of foliar spraying with Zn and different levels of phosphate fertilizer on growth and yield of sunflower plants grown under saline conditions, Egypt. J. Agron. 27 (2005) 11—22.

[65] M.E. Gobarah, M.H. Mohamed, M.M. Tawfik, Effect of phosphorus fertilizer and foliar spraying with zinc on growth, yield and quality of groundnut under reclaimed sandy soils, J. Appl. Sci. Res. 2 (8) (2006) 491−496.

[66] I. Horesh, Y. Levy, Response of iron-deficient citrus trees to foliar iron sprays with a low-surface-tension surfactant, Sci. Hortic. 15 (3) (1981) 227−233.

[67] D.J. Halliday, M.E. Trenkel (Eds.), IFA World Fertilizer Use Manual, 1992.

[68] R.C. Dixon, Foliar fertilization improves nutrient use efficiency, Fluid J. 11 (2003) 11−12.

[69] M.J. Conway, Aerial application of phosphate fertilisers to radiata pine forests in New Zealand, Emp. For. Rev. (1962) 234−245.

[70] J. Hagin, A. Lowengart, Fertigation for minimising environmental pollution by fertilizers, in: C. Rodriguez-Barrueco (Ed.), Fertilizers and Environment, Kluwer Academic Publishers, The Netherlands, 1996, pp. 23−25.

[71] V. Gowariker, V.N. Krishnamurthy, S. Gowariker, M. Dhanorkar, K. Paranjape, The Fertilizer Encyclopedia, John Wiley & Sons, 2009.

[72] S.S. Hebbar, B.K. Ramachandrappa, H.V. Nanjappa, M. Prabhakar, Studies on NPK drip fertigation in field grown tomato (Lycopersiconesculentum Mill.), Eur. J. Agron. 21 (1) (2004) 117−127.

[73] F.G. Viets, R.P. Humbert, C.E. Nelson, Fertilizers in Relation to Irrigation Practices, Irrigation of agricultural lands, 1967, pp. 1009−1023 (irrigationofagr).

[74] E.D. Threadgill, Chemigation via sprinkler irrigation: current status and future development, Appl. Eng. Agric. 1 (1) (1985) 16−23.

[75] P.M. Glibert, J. Harrison, C. Heil, S. Seitzinger, Escalating worldwide use of urea−a global change contributing to coastal eutrophication, Biogeochemistry 77 (3) (2006) 441−463.

[76] R.W. Howarth, Coastal nitrogen pollution: a review of sources and trends globally and regionally, Harmful Algae 8 (1) (2008) 14−20.

[77] X. Yang, S. Fang, Practices, perceptions, and implications of fertilizer use in East-Central China, Ambio 44 (7) (2015) 647−652.

[78] D. Hite, D. Hudson, W. Intarapapong, Willingness to pay for water quality improvements: the case of precision application technology, J. Agric. Resour. Econ. 27 (2002) 433−449, 1835−2016-148791.

[79] A.M. Adrian, S.H. Norwood, P.L. Mask, Producers' perceptions and attitudes toward precision agriculture technologies, Comput. Electron. Agric. 48 (3) (2005) 256−271.

[80] J.K. Ladha, H. Pathak, T.J. Krupnik, J. Six, C. van Kessel, Efficiency of fertilizer nitrogen in cereal production: retrospects and prospects, Adv. Agron. 87 (2005) 85−156.

[81] H.Y. Han, L.G. Zhao, Farmers' character and behavior of fertilizer application-evidence from a survey of Xinxiang County, Henan Province, China, Agric. Sci. China 8 (10) (2009) 1238−1245.

[82] Z. Cui, X. Chen, F. Zhang, Current nitrogen management status and measures to improve the intensive wheat−maize system in China, Ambio 39 (5−6) (2010) 376−384.

[83] Y.S. Tey, M. Brindal, Factors influencing the adoption of precision agricultural technologies: a review for policy implications, Precis. Agric. 13 (6) (2012) 713−730.

[84] M. Yang, Y. Fang, D. Sun, Y. Shi, Efficiency of two nitrification inhibitors (dicyandiamide and 3, 4-dimethypyrazole phosphate) on soil nitrogen transformations and plant productivity: a meta-analysis, Sci. Rep. 6 (2016) 22075.

CHAPTER 2

Fate of the conventional fertilizers in environment

Ashitha A.[1], Rakhimol K.R.[2], Jyothis Mathew[1]
[1]School of Biosciences, Mahatma Gandhi University, Kottayam, Kerala, India; [2]International and Inter University Centre for Nanoscience and Nanotechnology, Mahatma Gandhi University, Kottayam, Kerala, India

1. Introduction

The soils naturally hold many nutrients such as nitrogen, potassium, phosphorous, and calcium. These nutrients help in plant growth. While the soil nutrients are short in supply or missing, it will cause the plants to suffer from nutrient deficiency and stops the growth. This is due to the effect of nutrient level on proper plant functioning and also in the plant products. So once the crops are harvested, the nutrient supply in the soil must be refilled; for that, the farmers add nutrients to their soils. There are varieties of sources by which nutrients are supplied to the soils to maintain soil fertility [1], which includes organic matter, chemical fertilizers, and even by some plants. From the start of the agriculture itself, farmers used fertilizers for a nutrient supplement.

The fertilizers are any organic or inorganic material of natural or synthetic origin that is added to agriculture fields to supply one or more plant nutrients necessary to the growth of plants. Fertilizers not only provide plant growth, but they do also give an extra boost when needed by providing additional nutrients. Fertilizers improve the natural fertility of the soil or restore the elements taken by the previous plants from the soil. The elements that plants require in large quantities are termed as macronutrients whereas elements that required in small quantities are referred to as micronutrients. A fertilizer that contains primary macronutrients such as nitrogen, phosphorous, and potassium is considered as a complete fertilizer [2]. All the plants use the same inorganic forms of fertilizer in the soil despite their source.

In ancient times for effective agriculture development, farmers mainly rely on organic fertilizers. Organic fertilizers include animal manure, compost, legumes, and crop residues [3]. The organic fertilizers reach the crop slowly as they were slowly released into the soil as materials decompose. To overcome this limitation in fertilization and also due to the

Controlled Release Fertilizers for Sustainable Agriculture
ISBN 978-0-12-819555-0
https://doi.org/10.1016/B978-0-12-819555-0.00002-9

unavailability of large amounts of organic nutrients, the conventional or synthetic fertilizers were introduced into the market. They are manmade products created using a manufactured process of nutrients from natural sources. The conventional fertilizers are produced as nutrients in a concentrated form that plants can easily utilize. One of the main reasons for farmers using conventional fertilizers was because it is more specific in application. The exact nutrient composition is known, and the fertilizer readily releases on the time it was given.

The use of fertilizer is expensive and also can harm the environment if not applied properly. The complex system of the biogeochemical cycle has often negatively affected by the application of fertilizers. If the fertilizer is not given in sufficient quantity, it will affect the productivity of the crop. If surplus fertilizer is added, excess nutrients will run off the fields and pollute the environment. To avoid possible negative effects on the environment, care should be taken to use the right amount.

2. Major components of fertilizer residues

An understanding of the chemical nature of fertilizers is essential because the presence and quantity of each component will affect the growth and yielding of the plant. Most of the plants need 18 elements for their growth and development. Three major elements, carbon, hydrogen, and oxygen are given to the plants by air and water. However, all other elements are acquired from the soil (Fig. 2.1). The absence of any element will reflect their growth and yield.

Major fertilizing elements are nitrogen, phosphorous, and potassium (N, P, K). There are two types of fertilizers, complete and incomplete fertilizers, depending on the presence or absence of three major elements. If N, P, and K are present in the fertilizer, it is said to be complete and the fertilizer with the absence of any major element is said to be incomplete. The blending of two or more incomplete fertilizers will result in complete fertilizer.

3. Side effects of fertilizer residues

Even though chemical fertilizers have a role in increasing plant nutrients in unfavorable weather conditions or during the additional need of plant nutrients, not all the nutrients in the applied fertilizer is taken up by the growing plant. The fate of remaining residues was important to consider as their effects on long-term intensive use are unfavorable to our ecosystem.

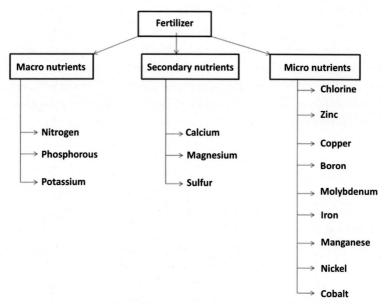

Figure 2.1 Essential nutrients of the plant.

These residues may remain in the soil; they may reach water bodies through the leaching process or running off the soil surface or may lose to the atmosphere due to volatilization. It can also cause chemical burns to crops, acidification of soil, and can lead to mineral depletion of the soil (Fig. 2.2). The intensity of negative effects depends on the ions involved in physicochemical and biological reactions.

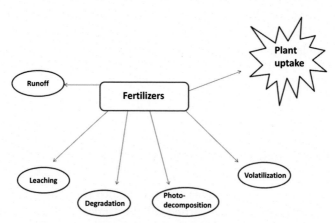

Figure 2.2 Major routes of applied fertilizers in fields.

3.1 Influence of mineral fertilizers

Low efficacy of nutrients assimilation in the environment causes serious problems in the aspects of environmental protection. It leads to the dispersion of excess biogenic compounds in the environment that were not completely used up by the plants. Thus the migration of these compounds becomes a worldwide problem. The consequences of dispersion affect the geological and geochemical processes of many ecosystems and last millions of years [1]. The mineral fertilizer compounds move from soil to water ecosystems as a result of areal runoffs, soil erosion, and rinsing out. This causes increased contamination and eutrophication of surface and sea waters.

The surplus amounts of fertilizers vary the degree of salinity and pH of the soil and lead to the gradual degradation of soil. The mineral fertilizers such as urea and ammonium salt are potent sources of ammonia and their migration to nearby ecosystems disturbs or destroys vegetation. A part of ammonia converts to nitric acid combines with sulfuric acid and form acid rain, which badly influences vegetation and leads to soil depletion and erosion. The nitrogen monoxide and nitric oxide are produced as a result of the denitrification process and cause the depletion of the ozone layer, subjecting people to the contact of ultraviolet radiation.

3.2 Phosphate anions

Phosphate is a nutrient ion that is particularly immobile in the soil. It is very strongly taken up by surface iron, aluminum, manganese oxides, and hydroxides and by clay particles. It can also be precipitated by calcium ions to produce calcium phosphates such as hydroxyapatite or in calcareous soils, calcium octaphosphate [4]. So if the soil is peat low in active iron or aluminum ions, or almost pure, then the phosphate portion of water-soluble phosphate fertilizer added to the soil will convert into forms that are low soluble in soil water. The phosphate fertilizer is not removed in the harvested crop, it remains; the remainings of the phosphate in the soil is washed off the land by runoff or by heavy rainfall. Except in the case of soil wash off the land by runoff or heavy rainfalls shortly after the phosphate fertilizer application to the wet soil surface, the phosphate remains in the soil and is not removed in harvested crops [4].

Most of the studies concluded that as phosphate is immobile in most soils, only a negligible amount of phosphate in fertilizer is leached from soils into drainage water. The phosphate is carried into rivers by soil erosion as the adsorbed soil particles are carried off the land and as soluble organic wastes through bodies of livestock and fodder plants.

3.3 Potassium cations

Most of the potassium fertilizers presently in use are soluble in water. However, once diffused in soil water, they adsorb to the soil particles. Potassium ions take part in cation exchange. When they are absorbed by a clay or humus particle, an equal amount of other cations are replaced into the soil. In normal soils, the replaced cations are chiefly calcium in addition to some magnesium. However, these ions are held less tightly than potassium, so the potassium ion concentration in soil solution usually ranges from 10^{-4} to 10^{-5} M. If the level of exchangeable potassium is increased through fertilizer addition, the proportion of exchangeable ions is converted to a form that is less accessible to the soil solution [4]. Thus the ions are said to be fixed and that takes place within the layers of micaceous and vermiculitic clay particles. Nearly all the soils have a great ability to hold potassium while soluble potassium salts are added to the soil because most soils are not composed of strongly weathered minerals containing those types of clay. Thus only a few proportions of fertilizer potassium might be lost by leaching due to soil stronghold for potassium.

3.4 Calcium and magnesium

As an increasing amount of nitrate leaching out of the soil is caused by the use of nitrogen fertilizers, this nitrate will be lost as calcium and magnesium nitrate. Thus to restore the removed calcium and magnesium in the soil and to prevent acidification, regular application of these limiting materials to calcareous soils is needed [4].

3.5 Nitrogen

Nitrogen is ubiquitous in the atmosphere and one of the most vital nutrients required for all living things to survive [5]. In the atmosphere, nitrogen accounts for 78% as elemental dinitrogen (N_2) gas. However, in the form of dinitrogen gas, N is inert and is not available directly for plant uptake and metabolism. As the soil-N supplies are inadequate for optimum growth production, most of the fertilizers applied to the soil contain nitrogen. The movement of soluble nitrogen from nitrogen fertilizers in the soil is complex. The regular nitrogen fertilizers in use are urea, ammonia, ammonium salts, nitrate salts, and also very limited use of proteinaceous residues such as bone waste and wool waste. When the well-drained soil is warm in normal farming conditions, all soluble nitrogen compounds are quickly oxidized to nitrate, thus not absorbed to soil particles.

The increase in the nitrogen levels in rivers and streams causes algae overgrowth. When the algae die and decompose, it leads to an increase in the organic matter in the water. This needs oxygen and thus the levels of oxygen in the water drop. The decrease in oxygen adversely affects the aquatic environment; fish, crabs and other aquatic life die as a result. The occurrence of blue-green algae blooms in the San Francisco Bay Delta is mainly due to the increased nitrogen levels. The algae bloom that appears in different colors also produces toxins that can harm aquatic life.

3.6 Phosphorus

The phosphate fertilizer added to the soil can influence soil acidity, mainly through the release or gain of H^+ ions by the phosphate molecule depending on soil pH. Monoammonium phosphate, single superphosphate, and triple superphosphate add phosphorus to soil in the form of $H_2PO_4^-$ ion that can acidify the soil with pH greater than 7.2 but in acidic soils that do not affect soil pH. The form of P in diammonium phosphate can make acidic soils more alkaline but has no effect on soil with a pH > 7.2.

Phosphorous will exist in the soil or sediment and can become mobile in wet, warm, and during waterlogging. Ordinary superphosphate can be washed deep into the soil before the annual plants have effectively established roots as it is 80% water-soluble. Absorption of P by crop has little effect on soil acidity due to the small amounts of P uptake in 1 year. Thus significant differences in rhizosphere pH were not observed for the absorption of different orthophosphate ions.

3.7 Sulfur

The sulfur fertilizer can also affect soil acidity when added to the soil, mostly through the release of H^+ ions by the addition of elemental S (S^0) or thiosulfate ($S_2O_3^{2-}$, in ammonium thiosulfate—ATS). However, when compared to the N uptake by the plants, the amount of S added to the soil and their uptake are normally small.

4. Effect on water pollution

Nitrogen from nitrogenous fertilizers in agricultural areas arrives at the water bodies in three ways: drainage, leaching [5] and flow. Nitrate is a basic component of fertilizers and also one of the significant parameters of water pollution [6]. Nitrate leaching is especially associated with fertilization in agricultural practices. In some of the irrigated agricultural land in arid and semiarid areas, along with the evaporation of water nitrate accumulation in the soil also increases. In soil, the nitrogen present in the fertilizer was converted into nitrate through nitrification by microorganisms. Nitrate reaches the groundwater owing to its negative charge. The most common form of dissolved nitrogen in groundwater is a nitrate, but also the form of nitrite (NO_2^-), nitrogen oxide (N_2O), nitrogen (N_2), and organic nitrogen can be found. The plants only utilize 50% of the nitrogenous fertilizers applied even in the ideal conditions. 2%−20% was lost by evaporation, 15%−25% interacts with the compounds in clay soil, and the leftover 2%−10% interferes with ground and surface water. The nitrogenous fertilizers enter surface and underground waters as they are not absorbed products.

The nitrate from the drinking water is absorbed by the body in the intestinal tract. Nitrate after entering the body is rapidly distributed throughout the tissues. Approximately 25% of nitrate ingested is secreted into saliva and is reduced to nitrite by oral microflora. The nitrate and nitrite swallowed then reenter the stomach [7]. Nitrate is a precursor for the formation of N-nitroso compounds and exposure to these compounds may result in serious health issues such as cancer, birth defects, and other adverse effects [8].

The most significant health problem associated with nitrate in drinking water is methemoglobinemia. According to the University of Nebraska−Lincoln Extension, the symptoms develop such as gastrointestinal swelling and irritation, diarrhea, and protein digestion problems. In the presence of nitrite, hemoglobin can be converted to methemoglobin that cannot carry oxygen. The enzymes in the blood of adults continually convert methemoglobin back to hemoglobin, thus normally do not exceed 1%. However, in newborn infants, the level of enzyme is lower and the level of methemoglobin is usually 1%−2%. Above these levels are considered as methemoglobinemia "blue baby syndrome" [1,6,9].

Eutrophication of the water is the major adverse effect of intensive fertilizer use [9]. The amount of algae formation and aquatic plants increases as a result of greater amounts of intensive fertilizer in water, thus causing loss of water quality and degradation of the water environment by eutrophication [4]. Eutrophication in the bottom layer of water bodies creates an oxygen-free environment. As a result, water becomes unsuitable for drinking and supply. It leads to a reduction in the number of living species in the aquatic environment, an increase of unwanted species and odor problems [10].

Some general approaches and specific ways to reduce the movement of ammonia to surface water or the movement of nitrate to groundwater are described as follows:

i. Reduce the load of total nitrogen
- Make sure the livestock feed measures are not higher than necessary to meet production targets.
- First use nitrogen from available sources on the farm, like manure, before buying any nitrogen fertilizers off-farm.

ii. Prevent runoff from nutrient materials or manure
- Proper storage of manure until the time of land application.

iii. Management of fields to avoid leaching to groundwater
- As sandy or gravelly soils and soils with shallow water are more susceptible to nitrogen leaching, the identification of fields sensitive to nitrogen is necessary.
- Nitrogen application should be matched with crop requirements.
- Nitrogen contribution from green manure crops and any previous crop rotation should be reported.
- Application of nitrogen just before the time of greatest crop uptake.
- Split applications of nitrogen through fertigation.
- To maintain soil health and to make efficient use of nitrogen, crop rotation should be practiced.
- To "tie up" any surplus nitrogen at the end of the season, the establishment of cover crops is needed.

iv. Management of nutrient application to avoid ammonium losses to surface water
- Practice timely preparation of land to incorporate manure, balancing the losses of nitrogen to the atmosphere.
- Application of manure near surface water or on steeply sloping land should be avoided.
- To prevent the runoff of fertilizers, low application rates should be kept.
- After application, mix the manure as soon as possible.
- To filter runoff before it enters surface water, use buffer strips and erosion control structures.

5. Effect on soil pollution

The effect of chemical fertilizers on the soil is not directly evident because of the strong buffering power due to the components of the soil.

Gradually the soil pollution causes deterioration of soil fertility, and the soil degradation reactions direct to the deterioration of the balance of the current elements. In addition, the toxic substances from the soil accumulate within the crops and cause adverse effects in humans and animals that fed.

In agriculture, soil structure has significance in productivity and is regarded as an indicator of soil fertility. High levels of potassium- and sodium-containing fertilizers have a negative impact on soil pH, soil structure deterioration [6]. Long-term use of acid-forming nitrogen fertilizers causes a decrease in soil pH and decline the efficacy of field crops [11].

Several actions can be taken to lower the acidification rates in soil, such as the following:

- Use less acidifying nitrogen fertilizers such as urea rather than ammonium sulfate.
- Saw early for maximizing the opportunity of the crop to recover soil nitrate.
- Use of deep-rooted crops.
- In the below root zone, minimize the water percolation.
- Must avoid extreme irrigation.
- Leave the manure where the animals graze or reduce the removal of manure from pastures.
- To minimize the unnecessary accumulation of soil organic matter under pasture, the use of cropping rotations is helpful.
- Feed hay on the field where it is cut.

6. Effect on air pollution

When fertilizers are applied excessively, it causes air pollution by nitrogen oxide (NO, N_2O, NO_2) emissions. The atmospheric gases such as carbon dioxide, methane, hydrogen sulfide (H_2S) with chloro-fluoro hydrocarbons, water vapour and halon gases are associated with these compounds [12]. In addition, the gases on lower layers of tropospheric ozone contribute to the greenhouse effects. Thus globally the atmospheric N_2O increases from 0.2% to 0.3% each year.

Ammonia is the main form of agricultural air pollution, which enters the air as a gas from heavily fertilized fields and livestock waste. Then it combines with pollutants mainly nitrogen oxides and sulfates from vehicles, power plants, and industrial processes to form tiny solid particles or aerosols, not more than 2.5 μm across. These particles can penetrate deep into the lungs, causing pulmonary or heart diseases. The fertilizer production will certainly keep growing to manage with food needs of human population. As a result, the amount of aerosols created depends on factors such as air temperature, precipitation, season, time of day, wind patterns, and other needed ingredients from industrial or natural sources.

7. Role of greenhouse gas emission

Around 10%—12% of total greenhouse gas (GHG) emissions are contributed by agriculture and are the main sources of noncarbon dioxide (CO_2) GHGs that emit nearly 60% of nitrous oxide and nearly 50% of methane. GHG emission from agriculture soils is due to the use of fertilizers in agriculture. A group of factors that determine the direct and indirect emission of GHGs from the agriculture soil is the rate of fertilizer and organic manure application, yield, and area under cultivation. The sources of direct emission include N fertilizers, crop residues, and the mineralization process of soil organic matters.

7.1 Nitrous oxide emissions

N_2O is produced from nitrogen in soils and animal waste by microbial transformation and thus frequently associated with N fertilizer inputs in agriculture [13]. In the case of overload of fertilizer, soil microbes on a farm may release high levels of nitrous oxide, which is 265 times as much heat-trapping power as carbon dioxide. It has been estimated that the N_2O emission from agricultural land will further enhance by 35%—60% by 2030 as the use of nitrogen fertilizer and manure production is increased [13].

7.2 Methane emission

An anthropogenic source for methane is the rice fields and they contribute up to 20% or ~ 100 Tg CH_4 globally on an annual basis. While comparing with that of CO_2 the global warming potential of methane is 28 times greater [13]. The reduction in redox potential after flooding favors methane production in soils. The application of organic and chemical fertilizers enhances methane production. The nitrogen fertilizers induce crop growth and supply more carbon substrates (through organic root exudates and sloughed-off cells) to methanogens for CH_4 production. The nitrogen fertilizers also alter the activities of methanotrophs in the soil. During greater availability of ammonium-N is available in soils, the methanotrophs switch substrates from CH_4 to ammonia.

8. Effect on food quality

The nitrogen fertilizers will notably raise the nutritional value of the grass for livestock feed. During the growing period, a high amount of fertilizer nitrogen can be successfully converted into grass protein. In the case of crops grown for human consumption, the application of fertilizer to well-farmed land has little effect on flavor and nutritional quality of crops. Because adding nitrogen fertilizer tends to elevate the protein content of the crop but may not raise the content all the essential amino acids in the same ratio. Similarly, the addition of phosphate fertilizer may raise the phosphate content of the crop, but the effect is normally little and rarely of economic significance.

Heavy nitrogen uses have disadvantages such as increase the liability of crop leaves to infect by a number of diseases and pests, increased tendency of cereals to lodge and may reduce the quality of the crop in an evident manner. The storage and cooking quality of potatoes were lowered by a high level of fertilizer nitrogen and in the case of sugarbeet, it will be difficult to refine sugar. However, it is profitable, the short-term overuse of nitrogen fertilizer for horticulturist growing salad and leafy vegetable crops, as the nitrogen gives rapid leaf growth and soft succulent leaves. Most of the fertilizers contain substances and chemicals including methane, carbon dioxide, ammonia, and nitrogen, which results in an increased quantity of GHGs. After carbon dioxide and methane, the third most major GHG is nitrous oxide, a byproduct of nitrogen destroys the ozone layer. The ozone layer protects the earth from harmful ultraviolet rays of the sun, so depletion of ozone causes global warming and subsequent weather changes. A vital ingredient in most of the fertilizers used is peat. Peat bogs tend to store a greater amount of carbon dioxide than the tropical forests of the entire world. Harvesting these could increase the concentration of carbon dioxide in the atmosphere and again worsen the problems.

8.1 Impact of fertilizer use on heavy metal accumulation in agricultural land

The continued use of agriculture fertilizer with the presence of metals could lead to the accumulation of these metals in soils to toxic levels. Even though the major plant nutrients in fertilizers are subjected to crop removal and leaching, heavy metals will continue in the soil for a much longer period of time. The enormous use of inorganic fertilizers globally is related to the accumulation of contaminants, for example, arsenic (As), cadmium

(Cd), fluorine (F), lead (Pd), and mercury (Hg) in agricultural soils. The pure phosphate fertilizers and high phosphorous blended fertilizers contain high levels of numerous elements of potential environmental concern. The As (0.3−16 mg/kg) and Cd (0−166 mg/kg) content is higher in phosphate fertilizer. The chances of heavy metal to affect human and animal health is higher, as the long period use of contaminated fertilizers for crop production may contaminate the soil and water resources and enters into the food chain through crop plants. Some of the studies had reported that chemical fertilizer usage for term increases the concentration of metals such as cadmium beyond the soil standard limit for agriculture purposes. The factors that depend on metal contamination of soils are the location of the soil, fertilizer source, and organic matter of the soil.

The radioactivity in fertilizers differs significantly depending on their concentrations in the parent mineral and on the fertilizer production process. The high rates of annual phosphorous fertilizer use results in uranium-238 concentrations in soil and drainage water, which are several times greater than normal ranges. Yet, the impact of radionuclide contaminated food on the risk to human health is very small.

9. Fertilizer guidelines

9.1 Superphosphate

- Do not apply fertilizer on the surface layers of dams, streams, or swampy areas.
- Do not apply fertilizer on bare ground.
- Groundcover around dams and streams should be properly maintained
- Avoid fertilizer application when heavy cyclonic rain is expected.

9.2 Nitrogen

- Affordable least acidifying fertilizer should be used.
- To minimize nitrate leaching, apply fertilizers in small amounts frequently and avoid application all at once.

9.3 Impact of fertilizers on human health

Conventional fertilizers release their contents into the surrounding within a short time after their application in soil. A small amount of these fertilizers are used by the plants. Above 50% of the applied fertilizers escape from the applied site and enter into the soil, water, and atmosphere. All the fertilizers

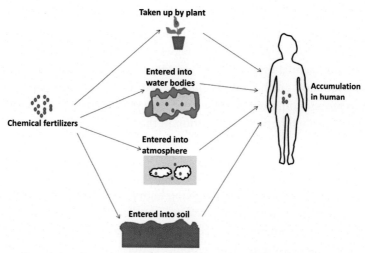

Figure 2.3 Various routes for accumulation of fertilizers in human.

will reach into the human body directly or indirectly. Accumulation of these fertilizer components will lead to major health problems in humans (Fig. 2.3).

Phosphate fertilizers are considered to possess an array of trace elements. Cadmium, uranium, chromium, molybdenum, mercury, and lead are the major toxic elements contained in phosphate fertilizers [14]. These elements will accumulate in the fields over the repeated application of fertilizers for many years. The accumulated elements will slowly be leached out from the soil and enter into the surface and groundwater. Oyedel et al. studied the heavy metal accumulation in vegetable tissues and they found that most of the heavy metals accumulate in the shoot and root of the plants. The largest accumulation was found for the elements, cadmium and lead [15].

Uranium is an abundant element in the environment. When it enters the human body, it will be absorbed from the stomach and enter into the bloodstream. The accumulation of uranium is mainly seen in kidney and bone. It will form a uranium compound and lead to the replacement of calcium from the bone by these compounds. The main health issue of uranium is nephritis [16–18]. Cadmium will enter into the body through the skin, gastrointestinal tract, and pulmonary tract. It is a nonessential element of the human body and it will cause kidney and lung disorders [19]. Lead and mercury are the other elements that cause a carcinogenic effect in humans [20].

10. Conclusion

The rapid increase in the global population demands a tremendous rise in food and agriculture yield, which increases the overall use of fertilizers. The fertilizers are essential for the proper growth and health of the plant and also for improved crop yield. Even though the fertilizers can boost plant growth and crop production, their excessive use have drastic side effects in the long run. The too much fertilizer application in the soil can alter soil fertility due to the increased acid levels of soil. So soil test done at least once in every 3 years is recommended to keep track of whether the right amount and required nutrient containing fertilizer is being used.

As government and society are giving high priority on natural resources, the environmental issues by fertilizer applications were also highly considered. Government and private sectors have funded considerable research for identifying and developing better management practices to minimize the pollution of the environment by excess fertilizer usages. Researchers and hydrologists are developing several methods to slow the rate of overland stormwater flow, allowing sediments to deposit onto the vegetation, thus no longer moving toward water bodies. The reduced flow rate can infiltrate the stormwater contributing groundwater recharge than entering to other water bodies.

References

[1] K. Lubkowski, Environmental impact of fertilizer use and slow release of mineral nutrients as a response to this challenge, Pol. J. Chem. Technol. 18 (1) (2016) 72–79.
[2] R. Kumar, R. Kumar, O. Prakash, Chapter-5 the impact of chemical fertilizerson our environment and ecosystem, Chief Editor 35 (n.d.) 69.
[3] G. Evanylo, C. Sherony, J. Spargo, D. Starner, M. Brosius, K. Haering, Soil and water environmental effects of fertilizer-, manure-, and compost-based fertility practices in an organic vegetable cropping system, Agric. Ecosyst. Environ. 127 (1–2) (2008) 50–58.
[4] R.A. Olsen, Effects of Intensive Fertilizer Use on the Human Environment, FAO, 1972.
[5] C.W. Liu, Y. Sung, B.C. Chen, H.Y. Lai, Effects of nitrogen fertilizers on the growth and nitrate content of lettuce (*Lactuca sativa* L.), Int. J. Environ. Res. Publ. Health 11 (4) (2014) 4427–4440.
[6] S. Savci, An agricultural pollutant: chemical fertilizer, Int. J. Environ. Sustain Dev. 3 (1) (2012) 73.
[7] World Health Organization, Nitrate and Nitrite in Drinking-Water: Background Document for Development of WHO Guidelines for Drinking-Water Quality (No. WHO/SDE/WSH/04.03/56), World Health Organization, 2003.
[8] M.H. Ward, R.R. Jones, J.D. Brender, T.M. De Kok, P.J. Weyer, B.T. Nolan, S.G. Van Breda, Drinking water nitrate and human health: an updated review, Int. J. Environ. Res. Publ. Health 15 (7) (2018) 1557.

[9] A.T. Ayoub, Fertilizers and the environment, Nutrient Cycl. Agroecosyst. 55 (2) (1999) 117–121.

[10] P.S. Chauhan, A. Singh, R.P. Singh, M.H. Ibrahim, Environmental impacts of organic fertilizer usage in agriculture, Org. Fertil. Types Prod Environ. Impact (2012) 63–84.

[11] P. Barak, B.O. Jobe, A.R. Krueger, L.A. Peterson, D.A. Laird, Effects of long-term soil acidification due to nitrogen fertilizer inputs in Wisconsin, Plant Soil 197 (1) (1997) 61–69.

[12] S. Savci, Investigation of effect of chemical fertilizers on environment, APCBEE Procedia 1 (2012) 287–292.

[13] S. Lenka, C.M. Rajendiran, M.L. Dotaniya, J.K. Saha, Impacts of fertilizers use on environmental quality, in: National Seminar on Environmental Concern for Fertilizer Use in Future at Bidhan Chandra Krishi Viswavidyalaya, Kalyani on February, vol. 26, 2016.

[14] G.H. McClellan, Mineralogy of carbonate fluorapatites, J. Geol. Soc. 137 (6) (1980) 675–681.

[15] D.J. Oyedele, C. Asonugho, O.O. Awotoye, Heavy metals in soil and accumulation by edible vegetables after phosphate fertilizer application, Electron. J. Environ. Agric. Food Chem. 5 (4) (2006) 1446–1453.

[16] Y.T. La, D.L. Willis, O.I. Dawydiak, Absorption and biokinetics of U in rats following an oral administration of uranyl nitrate solution, Health Phys. 53 (2) (1987) 147–162.

[17] I.M. Fisenne, P.M. Perry, Isotopic U concentration in human blood from New York City donors, Health Phys. 49 (6) (1985) 1272.

[18] P. Kurttio, A. Auvinen, L. Salonen, H. Saha, J. Pekkanen, I. Mäkeläinen, H. Komulainen, Renal effects of uranium in drinking water, Environ. Health Perspect. 110 (4) (2002) 337–342.

[19] J. Godt, F. Scheidig, C. Grosse-Siestrup, V. Esche, P. Brandenburg, A. Reich, D.A. Groneberg, The toxicity of cadmium and resulting hazards for human health, J. Occup. Med. Toxicol. 1 (1) (2006) 22.

[20] H. Savoy, Fertilizers and Their Use. PB-1637, Agricultural Extension Service, The University of Tennessee, Knoxville, 1999.

CHAPTER 3

Controlled release of fertilizers—concept, reality, and mechanism

Maya Rajan, S. Shahena, Vinaya Chandran, Linu Mathew
School of Biosciences, Mahatma Gandhi University, Kottayam, Kerala, India

1. Introduction

A fertilizer is any substance or material that has either natural or synthetic origin and is distributed to the soil or plants to supply plant nutrients that are essential to the growth of plant tissues. Chemical fertilizers offer mainly three important macronutrients, namely phosphorous, potassium, and nitrogen. Fertilizers may also add some secondary nutrients such as sulfur, calcium, and magnesium to the media or soil. Different types of fertilizers are used depending on the need of the crop for growth promotion. Chemical fertilizers are made with artificial or synthetic ingredients. However, they are more expensive than natural fertilizers, and also they may contain certain toxic materials that cause allergies. The excess or uncontrolled use of fertilizers destroys the soil properties, such as osmolarity and water holding capacity. Excess usage can also adversely affect soil microorganisms.

Some weed plants can also act as fertilizers. For example, chickweed, yellow dock, etc. are used as green manures because of their high nitrogen content. The decomposed organic matter also acts as natural fertilizers. Organic compost or well-aged herbivore manure creates nutrient-rich organic material capable of improving soil quality and texture.

1.1 Controlled release fertilizers

Effectiveness of nutrient supply and increased nutrient use efficiency with reduced environmental pollution depends on two factors: matching nutrient supply with plant demand and maintaining nutrient availability. One effective method for increasing nutrient use efficiency and minimizing environmental hazards is the use of controlled release fertilizers (CRFs). The term CRFs refers to fertilizer granules that intercalate within carrier

Controlled Release Fertilizers for Sustainable Agriculture
ISBN 978-0-12-819555-0
https://doi.org/10.1016/B978-0-12-819555-0.00003-0

molecules and thereby improve the efficiency of nutrient release to the crops and reduce the ecological, environmental, and health hazards [1]. Fertilizers encapsulated in organic or inorganic material have a controlled rate and duration of plant nutrient release. One of such CRFs is polymer-coated urea [2,3]. CRFs provide a nutrient release for a longer period as compared to quick release fertilizers. CRFs increased the availability of nutrients due to the controlled release of nutrients into the fixing medium during the fixation process in the soil [4]. They increase the efficiency of nutrient availability and reduce the environmental degradation. In developed countries, fertilizer usage has increased by about 5% in each year from 1950 to 1980 but expects to remain more or less constant in the future [5]. Generally, CRFs are coated with polymers for their controlled release. Polymer can be either natural or synthetic and the nutrients are released from the polymer at a controlled rate. These polymers are biodegradable or nonbiodegradable. Polyvinyl chloride, polyacrylamide, and rubber are examples of synthetic biodegradable polymers. One of the advantages of these biodegradable polymers is the lack of toxicity and biological activity. Control release or slow release fertilizers potentially minimize environmental hazards and may improve the nutrient use efficiency [6]. This technology is generally applied to those fertilizers in which the factors dominating the rate, pattern, and duration of release are well known and are controllable during the preparation of controlled release fertilizers [7,8]. The terms controlled release and slow release fertilizers are used interchangeably yet they are different. However, the actual difference between these two is not very clear [7].

1.2 Slow release fertilizers

Slow release fertilizers are the fertilizers characterized by the release of nutrients at a slower rate but the factors such as rate, pattern, and duration of release are not well controlled and may be strongly affected by handling conditions such as storage, transportation, distribution in the field and soil conditions such as moisture content and biological activity [9,10]. In slow release, the nutrient release depends on soil and climatic conditions that cannot be predicted. However, in CRFs, the release rate, time, and quantity can be predicted within certain limits [11]. Nitrogen compounds decomposed by microorganisms are commonly called as slow release fertilizers (SRFs). In SRFs, the release of nutrients completely depends on soil and climatic conditions. It releases the nutrients gradually with time and it might be in organic or inorganic form. Nitroform is an example of inorganic SRFs [3]. Urea-formaldehyde (UF) and urea-isobutyraldehyde (IBDU) are organic SRFs [7]. In addition, natural and synthetic SRFs are available. Natural SRFs are plant manure, animal manure, and compost [12]. Because of the organic nature of these compounds, microorganisms are allowed to decompose these components before they release nutrients into the crops. These fertilizers may take more time to release nutrients and may depend on soil microbial activity, which is influenced by soil moisture and temperature. Organic SRFs contain both macro- and micronutrients. Macronutrients are nitrogen, phosphorous, and potassium. Iron, manganese, and copper are the micronutrients.

Synthetic SRFs are partially water-soluble. In their case also, the availability depends on soil conditions. It contains a single nutrient at a higher level than that of natural SRFs. N-Sure is one of the examples that contains about 28% nitrogen [13].

1.3 Control release fertilizers versus slow release fertilizers

Controlled release fertilizer is also known as delayed release fertilizer or controlled availability fertilizer or coated fertilizer [14] or slow acting fertilizer [15]. In CRFs, the factors dominating the release of nutrients such as rate, duration of release, and pattern are controllable and well known during CRF preparation. In SRFs, the nutrient release rate is slower than water-soluble fertilizers but their pattern and duration of release are not controllable. SRFs always depend on microbial action that depends on soil conditions.

2. Factors influencing fertilizer release

One of the important factors for the release of nutrients is moisture content. When the growing medium dries off, it can greatly reduce the fertilizer release. It works well at normal temperature in case of crops. At temperatures below 10°C, there is no fertilizer release. If the temperature is above 32°C, the fertilizer release is excessive.

3. How does it work?

It is applied as either preincorporated into the growing medium or applied as a top dress before planting. One disadvantage of this is the prills fall out of the container when it is disturbed. Another method is mixing controlled release fertilizer directly into the growing medium. In this case, prills remain in the medium.

The working principle of controlled release fertilizers is that every granule is covered by an organic coating controlling the release of nutrients. These granules contain NPK, Mg, B, Cu, Fe, Mo, and Zn. Water that penetrates through the coating dissolves the nutrients in the granules. These dissolved nutrients are released from the granules by osmosis. Depending on the temperature, these nutrients are released at a constant and controlled rate.

4. Different types of slow or controlled release fertilizers [16]

1. Sulfur-coated urea
2. Urea formaldehyde
3. Resin-coated fertilizer
4. Urea and nitrification inhibitors
5. Sulfur-coated compound fertilizers

The use of CRFs is associated with several economic, agronomical, and environmental benefits. Economically, it reduces the cost, by ensuring the supply of nutrients to the crops for the entire season through a single application thereby reducing the demand for short-season manual labors [4]. Agronomically, the usage of CRFs is associated with enhancement of plant growth factors such as it reduces the stress and increases the availability of nutrients due to controlled release. In environmental aspects, the CRFs increased nutrient utilization efficiency, thereby reducing the loss of excess nutrients into the environment. High accumulation of fertilizer in the environment is minimized and, as a result, it decreases environmental problems associated with conventional fertilizer use such as eutrophication, which causes the water oxygen depletion, death of fishes, unpleasant odor, etc [9,17].

5. Advantages of CRFs

- They reduce the toxicity caused through high soil ionic concentration from the quick dissolution of conventional soluble fertilizer, that is, in some cases from ammonia or after the application of urea. Hence, CRFs improve the agronomic safety [18].
- Due to the reduction of toxicity and salt content of the substrates, they allow the application of substantially larger amounts of fertilizer as compared to conventional soluble fertilizers. This may reduce cost and save labor, time, and energy.

- They reduce the losses of nutrients especially losses of nitrate nitrogen by the uptake of nutrients by the plants through gradual nutrient release. They also reduce the evaporation losses of ammonia. This allows decreasing the risk of environmental pollution [19–21].

6. Disadvantages of CRFs

- Manufacturing of CRF is costlier than conventional fertilizers.
- Sulfur-coated urea always lowers the soil pH. Acidification of soil creates nutrient disorders resulting in calcium and magnesium deficiencies.
- CRF may be inadequate sources of nutrients in situations with low ambient and soil temperature.

7. Classification of controlled release fertilizers

Controlled release fertilizers are mainly classified into two:

7.1 Slow versus controlled release

The term simply implies that the nutrient released into the environment occurs in a more or less slow and controlled manner. SRFs release the nutrient in a slower manner compared to common fertilizer. The distribution rate and duration of release are not controlled. However, in CRFs, their duration of release, pattern, and rate are well known and also controllable during CRF preparation [18]. In the case of nitrogen fertilizers, ammonium and urea are stabilized by inhibitors that may be classified as "slow acting nitrogen." These fertilizers may be very effective in soil having a medium-to-high cation exchange capacity and good storage capacity of ammonium and other nutrients [22].

7.2 Systematic classification

Controlled release fertilizers are commonly classified into three types:

7.2.1 Organic N low solubility compounds

They are further classified into biologically decomposing compounds based on urea aldehyde condensation products such as urea formaldehyde and chemically decomposing compounds such as isobutyledene diurea (IBDU).

7.2.2 Fertilizers in which a physical barrier controls the release

Fertilizers that can appear as a core or granules that are covered with hydrophobic polymers or matrices in which the active soluble material is dispersed restrict the dissolution of the fertilizer. The coated fertilizers are of different types: fertilizers coated with an organic coating (resins) and fertilizers coated with inorganic materials (sulfur- or mineral-based coating).

7.2.3 Inorganic low solubility compounds

These type of fertilizers contain ammonium phosphate and partially acidulated rock phosphate.

8. Controlled and slow release fertilizers

8.1 Slowly releasing organic N compound

8.1.1 Urea formaldehyde condensation products

Condensation of urea with aldehyde is one of the most common methods for the preparation of CRFs. Urea formaldehyde is the organic N compound mainly used for the slow release of nitrogen [23]. It is prepared by the reaction of excess urea under controlled pH and temperature. This product contains unreacted urea, dimers, and oligomers. The addition of acids to this mixture results in the production of longer chain and water-soluble oligomers that are also slow in releasing nitrogen [24].

Based on the release of nitrogen from these UF products, the compounds are again grouped into three fractions:

1. Cold water-soluble N—it contains mainly urea, dimmers, soluble short UF chains. In this fraction, N is readily available.
2. Hot water-soluble N—it contains methylene urea and chains of intermediate length. N is slowly released into the soil.
3. Hot water-insoluble N—it contains intermediates and long chain and extremely slow decomposing or practically unavailable nitrogen.

The decomposition of UF is mainly due to the microbial action. So the release of nitrogen from these compounds mainly depends on soil properties such as biological activity, pH, moisture content, and temperature [24—26].

8.1.2 Synthetic nitrogen compounds

Isobutylidene diurea is prepared by reacting liquid isobutaldehyde with solid urea. It contains 31% of N. The release of N is by hydrolysis and depends indirectly on the particle size of the granular material and directly on moisture content of the soil. It also depends on microbial activity and occurs at low temperature [25]. Another example of synthetic nitrogen compounds is urea-triazone solution formed by the reaction of urea—ammonia—formaldehyde. It contains 28% nitrogen.

8.2 Coated fertilizers

8.2.1 Nonorganic coatings

One of the examples for nonorganic coating fertilizer is sulfur-coated urea. It is prepared by coating preheated urea granules with molten sulfur. This product contains about 31%—38% N. To reduce the microbial contamination, the wax sealant is sprayed to seal the crack [27—29]. The release of N from sulfur-coated urea depends on the quality of the coating. Sulfur-coated urea granules consist of three types of coatings such as damaged coatings with cracks, damaged coatings whose cracks were sealed with wax and perfect and thick coatings. Sulfur-coated urea coated with a damaged coating releases the urea immediately when brought in contact with water.

8.2.2 Polymer coating of sulfur-coated fertilizers

One of the CRF products that are manufactured by sulfur-coated fertilizer is again coated by a thin layer of the organic polymer (PSCU). This polymer layer provides resistance to coated granules. The release is the same as that of polymer-coated controlled release of fertilizers.

8.2.3 Fertilizers coated with organic polymers
8.2.3.1 Resin-coated fertilizers

Coatings are commonly prepared by in situ polymerization of cross-linked hydrophobic polymer that usually degrades upon heating. The most commonly used resin is alkyd-type resin (osmocote) and polyurethane-type coating [30]. The first resin-coated CRF is osmocote produced in California. The release of nutrients of these depends on coating composition and thickness [31]. In the case of osmocote, water penetrates through microscopic pores and creates osmotic pressure within the coated core. As a result, the coating stretches and the micro-pore size increases and the nutrients are released [32].

8.2.3.2 Thermoplastic polymer-coated fertilizers

This technology provides coating granular fertilizer with thermoplastic material, that is, polyethylene. The coating material is dissolved in a chlorinated hydrocarbon and sprayed it on the granules in a fluidized bed reactor [33,34]. Controlled release is achieved by blending the low permeability polyethylene with high permeability polymer such as ethylene vinyl acetate.

8.2.3.3 Matrix-based slow release fertilizers

To prepare these types of CRFs, the nutrients are mixed with materials that lowered the dissolution rate. Different types of materials were used for this purpose such as rubber [35], thermoplastic polymers, and gel-based matrix [36]. Other types of matrix-based fertilizers are Tablets and Spikes. In this case, plant nutrients are mixed and compacted with a binder [32,37]. It is mainly used in home gardening.

8.2.3.4 Low solubility inorganic fertilizers

A group of compounds having common formula $MeNH_4PO_4 \times H_2O$ are commonly used as low solubility fertilizers. "Me" stands for divalent cation (Mg, Fe, and Zn). Potassium analogs of ammonium salts are also prepared. For slow release of nitrogen, phosphorous, and potassium, fertilizer mixture of potassium and ammonium salts are also used [32,38,39]. Partially acidulated rock phosphate can be considered as slow release phosphate fertilizers, which are mainly used in acidic soil and light-textured soil [40,41]. If the soil has high phosphate fixation, these elements are readily transferred into low solubility form immediately after the application.

9. Mechanisms of control release

According to the mode of release control, four types of release are proposed: (1) diffusion, (2) chemical reaction or decomposition, (3) swelling, and (4) osmosis [42]. The fertilizer encapsulated in a biodegradable polymer in which water is diffused nutrients will become soluble. This is followed by the penetration of the microorganisms and degradation of the insoluble part of the fertilizer. Several factors are affecting the decomposition of biodegradable polymers such as morphology, molecular weight distribution, and chemical and physiochemical factors such as pH, temperature, and mechanical strength [43]. Nutrients are released from the biodegradable polymer by hydrolysis of the polymeric chain into nontoxic smaller molecules.

CRFs provide a good tool to control the nutrient release into the soil and to match the plant demand. They are expected to provide high use efficiency and reduce harm to the environment. The proper release of nutrients and proper utilization require tools for predicting the release under different soils and varying environmental conditions. The release curve of chemical fertilizers is characterized by too high initial release that is "burst" and too slow release of about last quarter to third of the nitrogen that is "tailing effect," which is significantly different from the pattern of plant uptake of the nutrients (sigmoidal form) [5,9,10]. Fertilizers coated with hydrophobic materials especially polymer coated provide a good control over the release. The release pattern from coated fertilizers is normally temporal release with or without burst that may follow the pattern of parabolic release through the linear pattern to sigmoidal release as the normal pattern of release of nutrients and plant uptake [44].

According to the multistage diffusion model, water penetrates the coating to condense on to the solid fertilizer core followed by the partial dissolution. An osmotic pressure will develop and granules swell leading to either of the two processes. In one process, this osmotic pressure causes coating burst and the entire fertilizer core is spontaneously released out. This process is called "failure mechanism or catastrophic release." In the second process, the membrane withstands the developing pressure and the core fertilizer is released slowly via diffusion: this is called diffusion mechanism [45]. The release of coated fertilizer is normally by the release of nutrients from fertilizer—polymer interface to polymer—soil interface and this is driven by water. Diffusion or swelling, degradation of the polymer coating, and dissolution are the parameters governing the releasing mechanism. Similar mechanisms were reported earlier [46—50].

Polymer degradation can take place either as a whole or at the surface, resulting in nutrient release at the delivery system. Erosion occurs at the surface by the hydrolytic or enzymatic reaction of the microorganisms at the surface of the coated granules. However, in the case of bulk erosion, water uptake followed by polymer degradation occurs by hydrolytic actions [46].

10. Factors affecting the nutrient release

1. Fertilizer type (solubility, density)
2. Thickener type
3. Thickener concentration
4. Size of device
5. Temperature of soil
6. Moisture in the soil

11. Mode of nutrient release from CRFs

11.1 Diffusion model

This is the first predicted model for the release of urea from sulfur-coated granules under soil conditions [29]. According to this model, the sulfur coating becomes cracked and after the granules are applied to the soil the coatings become degraded due to the microbial action. As a result, pores become exposed and allow water to enter. The urea diffuses from the granules through the pores by the erosion of the coating. According to Fick's first law,

$$dm/dt = -DS_K \frac{dC_K}{d\chi k}$$

where "M" is the mass of urea diffusing out of the granule, "D" is the diffusion coefficient of urea in water, "S_k" is the cross-sectional area through which diffusion occurs, and "C_k" is the urea concentration.

This model is unfortunately failed due to several factors relevant to diffusion from membrane-coated granules, that is, diffusion of urea from sulfur coating occurs in either in a steady N release phase during which urea dissolves or in a reduced rate phase where urea release rate decreases as the granule slowly empties. A model called the Arrhenius type of model for the diffusion coefficient D is proposed:

$$D = A T \exp(-2135/T)$$

where "T" is the absolute temperature in K, the value of 2135 (k^{-1}) stands for apparent energy of activation for urea diffusion [29].

Another model for the release of nutrients from coated granules was reported [51]. According to which diffusion coefficient is time dependent.

$$D = D_0 \cdot t^n$$

where t is the time, D_0 is an initial value at t = 0, and n is an empirical constant.

11.2 Empirical and semiempirical models

The model proposed that nutrient release is to be considered as a first-order decay process [52,53]. This model proposed two stages:
1. Water diffusion into the granule.
2. Solution flow out of the coating.

The model was then reduced to one equation, which was considered to describe only the second stage:

$$\frac{\log (Q_0 - Q_t)}{Q_0} = -kt$$

where Q_0 is the amount of fertilizer applied to the soil, Q_t is the cumulative quantity released after time t, and k is the decay rate constant. The release of nutrients depends on temperature and moisture. The decay rate constant k is linearly related to the water vapor pressure:

$$K = AP_w + B$$

P_w is the water vapor pressure at a given temperature.

11.3 Conceptual model

This model describes the different stages involved in the release from coated fertilizers. In the first stage, water penetrates through the coating and the vapor condenses in the solid core and creates internal pressure. In this stage, two pathways are possible. In one pathway, the internal pressure exceeds the membrane resistance and results in the rupturing of the core and the entire content of the granule is released immediately. This is known as failure mechanisms or catastrophic release [54]. In the second stage, the membrane resists the internal pressure and the fertilizer is released by diffusion driven by concentration gradient across the coating or by mass flow driven by pressure gradient or the combination of these two processes. This is known as diffusion. Failure mechanism was observed with nonelastic coatings such as sulfur or other inorganic coatings. Diffusion mechanism was observed with polymer-coated fertilizer such as polyurethane like coatings.

11.4 Diffusion release

According to this model, the release rate consists of three stages: (1) initial stage where there is no release is observed (lag phase), (2) constant release stage, and (3) gradual decay of release rate stage [55].

In the lag period, water vapor penetrates the granule. Fertilizer release is driven by a pressure gradient across the coating. The weight of the granule increases slightly. In the constant release stage, the fertilizer release starts when the critical volume of a saturated solution is formed; this also creates internal pressure. The release rate remains constant. The constant concentration provides a constant driving force for fertilizer transport. The volume present in this granule is always constant at this stage. The volume evacuated by the released fertilizer is occupied by water that continues to enter in the granule. Once the fertilizer in the core is dissolved, the internal pressure decreases and this is the decay stage. This part may be too slow, that is, a kind of moderate "tailing effect" is observed [10,56].

11.5 Failure release

In this mechanism, the fertilizer is released by the penetration of water through the polymer/ sulfur coated urea granules. The rate of water penetration depends on the driving force, the thickness of the coating and the coating material. The water vapor condenses in the granules and dissolves the fertilizer creating an internal pressure. The increased pressure causes the rupture of coating; as a result, fertilizer release will occur [55].

12. Application of CRFs in agriculture

Crops requiring a large amount of nutrients are generally applied with CRFs and SRFs [57]. Vegetables such as pepper [58], tomatoes [59], and onion [60] and fruits such as strawberries [61] and melon [62] are some of the examples. One single application with reduced labor cost plays an important role in convincing farmers to apply CRFs. In rice, increasing nutrient utilization efficiency requires several management approaches and the use of controlled release nitrogen was found to be very effective [32]. In Japanese rice fields, polymer-coated urea is used as CRFs to increase the efficiency of nutrients in rice [44]. Fruits such as banana and kiwi are some other examples where the CRFs are effectively used. These crops have high nutrient demand and are also high-income crops. Florida and California (United States of America) are the main users of CRF/SRFs for citrus farming [63].

13. Future aspects

- Development of soil degradable coatings
- Improved economic advantages
- Better assessment of expected benefits to the environment by controlled release fertilizers.
- Improved utilization of advanced technologies to prepare controlled release fertilizers.

14. Conclusion

Controlled release fertilizer releases the nutrients at a controlled rate or pattern into the soil. It is most commonly used in nurseries to fertilize trees and shrubs. The nutrient supply in a controlled manner increases the nutrient use efficiency and reduces environmental hazards. It reduces manual labor demands during critical periods of crop growth and reduces stress and toxicity. Temperature, soil fumigation, and moisture content are the important factors for the release of the nutrients from granules. The efficient, reliable, and cost-effective controlled release fertilizer formulation minimizes food crises and other challenges of crop production. When compared to conventional fertilizers, nutrient release from controlled release fertilizers minimizes leaching and increases fertilizer-use efficiency.

References

[1] C.V. Subbarao, G. Kartheek, D. Sirisha, Slow release of potash fertilizer through polymer coating, Int. J. Appl. Sci. Eng. 11 (1) (2013) 25—30.
[2] C.W. Du, J.M. Zhou, A. Shaviv, Release characteristics of nutrients from polymer-coated compound controlled release fertilizers, J. Polym. Environ. 14 (3) (2006) 223—230.
[3] S. Loper A.L. Shober, Soils & Fertilizers for Master Gardeners: Glossary of Soil and Fertilizer Terms1.
[4] A. Shaviv, Advances in controlled-release fertilizers, Adv. Agron. 71 (2001) 1—49.
[5] O.C. Bøckman, O. Kaarstad, O.H. Lie, I. Richards, Agriculture and Fertilizers, Norsk Hydro AS, 1990, p. 245.
[6] A. Alexander, H.U. Helm, Ureaform as a slow release fertilizer: a review, Z. für Pflanzenernährung Bodenkunde 153 (4) (1990) 249—255.
[7] M.E. Trenkel, Slow-and Controlled-Release and Stabilized Fertilizers: an Option for Enhancing Nutrient Use Efficiency in Agriculture, IFA, International Fertilizer Industry Association, 2010.
[8] D. Chen, H. Suter, A. Islam, R. Edis, J.R. Freney, C.N. Walker, Prospects of improving efficiency of fertiliser nitrogen in Australian agriculture: a review of enhanced efficiency fertilisers, Soil Res. 46 (4) (2008) 289—301.
[9] A. Shaviv, Plant response and environmental aspects as affected by rate and pattern of nitrogen release from controlled release N fertilizers, in: Progress in Nitrogen Cycling Studies, 1996, pp. 285—291.
[10] S. Raban, E. Zeidel, A. Shaviv, Release mechanisms controlled release fertilizers in practical use, in: J.J. Mortwedt, A. Shaviv (Eds.), Third Int. Dahlia Greidinger Sym. On Fertilisation And the Environment, 1997, pp. 287—295.
[11] M.C. Cartagena, J.A. Díez López, A. Vallejo, S. Jiménez, Evaluation and Classification of Coated Slow-Release Nitrogen Fertilizers by Means of Electroultrafiltration in an Integrated System, 1993.
[12] S. Shukla, E.A. Hanlon, F.H. Jaber, P.J. Stoffella, T.A. Obreza, M. Ozores-Hampton, Groundwater Nitrogen: Behavior in Flatwoods and Gravel Soils Using Organic Amendments for Vegetable Production, University of Florida Extension Service, 2006. Pub# CIR 1494.

[13] J.G. Clapp, Foliar application of liquid urea-triazone-based nitrogen fertilizers and crop safety, HortTechnology 3 (4) (1993) 442—444.

[14] J.J. Oertli, O.R. Lunt, Controlled release of fertilizer minerals by incapsulating membranes: I. Factors influencing the rate of release 1, Soil Sci. Soc. Am. J. 26 (6) (1962) 579—583.

[15] E.G. Gregorich, L.W. Turchenek, M.R. Carter, D.A. Angers, Soil and Environmental Science Dictionary, vol. 1, CRC Press, 2001, p. 22.

[16] S.I. Sempeho, H.T. Kim, E. Mubofu, A. Hilonga, Meticulous overview on the controlled release fertilizers, Adv. Chem. 2014 (2014).

[17] A.N. Sharpley, R.G. Menzel, The impact of soil and fertilizer phosphorus on the environment, Adv. Agron. 41 (1987) 297—324. Academic Press.

[18] D.J. Soldat, A.M. Petrovic, J. Barlow, Turfgrass response to nitrogen sources with varying nitrogen release rates, in: II International Conference on Turfgrass Science and Management for Sports Fields, vol. 783, June 2007, pp. 453—462.

[19] K. Lubkowski, B. Grzmil, Controlled release fertilizers, Pol. J. Chem. Technol. 9 (4) (2007) 83—84.

[20] F.L. Wang, A.K. Alva, Leaching of nitrogen from slow-release urea sources in sandy soils, Soil Sci. Soc. Am. J. 60 (5) (1996) 1454—1458.

[21] E. Rietze, W. Seidel, Vollbevorratung mitumhü lltem Langzeitdü nger senkt die Nitratauswaschung [An adequate supply of coated slowacting fertilizers reduces nitrateleaching], Versuchsstation Christinenthal (Germany), Urania Agrochem, Gartenbau Magazin. 3 (7) (1994) 32—33.

[22] A. Amberger, Research on dicyandiamide as a nitrification inhibitor and future outlook, Commun. Soil Sci. Plant Anal. 20 (19—20) (1989) 1933—1955.

[23] A. Shaviv, Preparation Methods and Release Mechanisms of Controlled Release Fertilisers, International Fertiliser Society, 1999.

[24] A. Alexander, H.U. Helm, Ureaform as a slow release fertilizer: a review, Z. für Pflanzenernährung Bodenkunde 153 (4) (1990) 249—255.

[25] H.M. Goertz, Commercial granular controlled release fertilizers for the specialty markets, in: R.M. Scheib (Ed.), Controlled Release Fertiliser Workshop Proceedings ", 1991, pp. 51—68.

[26] T. Aarnio, P.J. Martikainen, Mineralization of C and N and nitrification in scots pine forest soil treated with nitrogen fertilizers containing different proportions of urea and its slow-releasing derivative, ureaformaldehyde, Soil Biol. Biochem. 27 (10) (1995) 1325—1331.

[27] S.E. Allen, C.M. Hunt, G.L. Terman, Nitrogen release from sulfur-coated urea, as affected by coating weight, placement and temperature 1, Agron. J. 63 (4) (1971) 529—533.

[28] J.J. Oertli, Effect of coating properties on the nitrogen release from sulfur-incapsulated urea, Agrochimica 18 (1973) 3—8.

[29] W.M. Jarrell, L. Boersma, Release of urea by granules of sulfur-coated urea 1, Soil Sci. Soc. Am. J. 44 (2) (1980) 418—422.

[30] R.E. Lamond, D.A. Whitney, J.S. Hickman, L.C. Bonczkowski, Nitrogen rate and placement for grain sorghum production in no-tillage systems, J. Prod. Agric. 4 (4) (1991) 531—535.

[31] J.M. Lambie, Sierra Chemical Co, Granular Fertilizer Composition Having Controlled Release and Process for the Preparation Thereof, U.S. Patent 4657576, 1987.

[32] R.D. Hauck, Slow-Release and Bioinhibitor-Amended Nitrogen Fertilizers, Fertilizer Technology and Use, (Fertilizertechn), 1985, pp. 293—322.

[33] T. Fujita, C. Takahashi, S. Yoshida, H. Shimizu, Coated Granular Fertilizer Capable of Controlling the Effect of Temperature upon Dissolution-Out rate, U.S. Patent No. 4369055, U.S. Patent and Trademark Office, Washington, DC, 1983.

[34] S. Shoji, Controlled Release Fertilizers with Polyolefin Resin Coating, Development, properties and utilization, Konno, 1992.

[35] C. Hepburn, R. Arizal, A controlled release urea fertiliser. I: the encapsulation of urea fertiliser by rubber: processing and vulcanisation procedures, Plast. Rubber Process. Appl. 12 (3) (1989) 129—134.

[36] R.L. Mikkelsen, Using hydrophilic polymers to control nutrient release, Fert. Res. 38 (1) (1994) 53—59.

[37] F.N. Wilson, Slow Release-Ttrue or False? : A Case for Control. Proceedings-The Fertiliser Society of London, 1988.

[38] S.P. Landels, US markets for controlled-release fertilizers: present size and value, projected demand, trends, and opportunities for new CRF products, in: Controlled Release Fertilizer Workshop, 1991, pp. 87—101.

[39] S.P. Landels, Controlled-Release Fertilizers: Supply and Demand Trends in US Nonfarm Markets, Publisher: SRI International, Menlo Park, CA, USA, 1994.

[40] N.S. Bolan, M.J. Hedley, P. Loganathan, Preparation, forms and properties of controlled-release phosphate fertilizers, Fert. Res. 35 (1—2) (1993) 13—24.

[41] J. Hagin, R. Harrison, Phosphate rocks and partially-acidulated phosphate rocks as controlled release P fertilizers, Fert. Res. 35 (1—2) (1993) 25—31.

[42] L.T. Fan, S.K. Singh, Controlled Release: A Quantitative Treatment, vol. 13, Springer Science & Business Media, 2012.

[43] M. Stevanovic, D. Uskokovic, Poly (lactide-co-glycolide)-based micro and nano-particles for the controlled drug delivery of vitamins, Curr. Nanosci. 5 (1) (2009) 1—14.

[44] S. Shoji, H. Kanno, Use of polyolefin-coated fertilizers for increasing fertilizer efficiency and reducing nitrate leaching and nitrous oxide emissions, Fert. Res. 39 (2) (1994) 147—152.

[45] S.M. Lu, S.L. Chang, W.Y. Ku, H.C. Chang, J.Y. Wang, D.J. Lee, Urea release rate from a scoop of coated pure urea beads: unified extreme analysis, J. Chin. Inst. Chem. Eng. 38 (3—4) (2007) 295—302.

[46] M. Guo, M. Liu, F. Zhan, L. Wu, Preparation and properties of a slow-release membrane-encapsulated urea fertilizer with superabsorbent and moisture preservation, Ind. Eng. Chem. Res. 44 (12) (2005) 4206—4211.

[47] R. Liang, M. Liu, L. Wu, Controlled release NPK compound fertilizer with the function of water retention, React. Funct. Polym. 67 (9) (2007) 769—779.

[48] L. Wu, M. Liu, Preparation and properties of chitosan-coated NPK compound fertilizer with controlled-release and water-retention, Carbohydr. Polym. 72 (2) (2008) 240—247.

[49] M. Guo, M. Liu, R. Liang, A. Niu, Granular urea-formaldehyde slow-release fertilizer with superabsorbent and moisture preservation, J. Appl. Polym. Sci. 99 (6) (2006) 3230—3235.

[50] W.M. Jarrell, L. Boersma, Model for the release of urea by granules of sulfur-coated urea applied to soil 1, Soil Sci. Soc. Am. J. 43 (5) (1979) 1044—1050.

[51] V. Glaser, P. Stajer, J. Vidensky, Simulace prubehu rozpousteni obalovanych prumyslovych hnojiv ve vode-II, Chem. Prumysl. 37 (62) (1987) 353—355.

[52] M. Kochba, S. Gambash, Y. Avnimelech, Studies on slow release fertilizers: 1. Effects of temperature, soil moisture, and water vapor pressure, Soil Sci. 149 (6) (1990) 339—343.

[53] S. Gambash, M. Kochba, Y. Avnimelech, Studies on slow-release fertilizers: II. A method for evaluation of nutrient release rate from slow-releasing fertilizers, Soil Sci. 150 (1) (1990) 446—450.

[54] H.M. Goertz, Technology developments in coated fertilizers, in: Proceedings: Dahlia Greidinger Memorial International Workshop on Controlled/Slow Release Fertilizers, Technion-Israel Institute of Technology, Haifa, March 1993, pp. 7—12.

[55] S. Raban, Release Mechanisms of Membrane Coated Fertilizers, Research thesis, Israel Institute of Technology, Haifa, Israel, 1994.

[56] S.P. Friedman, Y. Mualem, Diffusion of fertilizers from controlled-release sources uniformly distributed in soil, Fert. Res. 39 (1) (1994) 19—30.

[57] S.P. Landels, Controlled-Release Fertilizers: Supply and Demand Trends in US Nonfarm Markets, Publisher: SRI International, Menlo Park, CA, USA, 1994.

[58] P.H. Everett, Controlled release fertilizers: effect of rates and placements on plant stand, early growth and fruit yield of peppers [Plastic mulch system], Proc. Fla. State Hortic. Soc. 90 (1977) 390—393.

[59] D.D. Gull, S.J. Locascio, S.R. Kostewicz, Composition of greenhouse tomatoes as affected by cultivar, production media and fertilizer, Proc. Fla. State Hortic. Soc. 90 (1977) 395—397.

[60] B.D. Brown, A.J. Hornbacher, D.V. Naylor, Sulfur-coated urea as a slow-release nitrogen source for onions, J. Am. Soc. Hortic. Sci. 113 (1988) 864—869.

[61] C. Cadahia, A. Masaguer, A. Vallejo, M.J. Sarro, J.M. Penalosa, Pre-plant slow-release fertilization of strawberry plants before fertigation, Fert. Res. 34 (3) (1993) 191—195.

[62] R.P. Wiedenfeld, Rate, timing andslow-release nitrogen fertilizers on bellpeppers and muskmelon, Hort. Science. 21 (1986) 233—235.

[63] F.L. Wang, A.K. Alva, Leaching of nitrogen from slow-release urea sources in sandy soils, Soil Sci. Soc. Am. J. 60 (5) (1996) 1454—1458.

CHAPTER 4

Characteristics and types of slow- and controlled-release fertilizers

Aiman E. Al-Rawajfeh[1], Mohammad R. Alrbaihat[2], Ehab M. AlShamaileh[3]

[1]Department of Chemical Engineering, Tafila Technical University, Tafila, Jordan; [2]Ministry of Education, Ajman, United Arab Emirates; [3]Department of Chemistry, The University of Jordan, Amman, Ajman, Jordan

1. Introduction

Agriculture is one of the main sectors of the world economy that plays a very crucial role as a food producer and as a place of employment for millions of people. Recently, the intensification of agricultural production was possible due to the use of high yielding varieties, irrigation, and mechanization, as well as soil feeding with mineral fertilizers and crop protection with pesticides [3,30]. Consumption of these agrochemicals has essentially increased during the last 50 years and resulted not only in the growth of agricultural production but also in the pollution of natural environment.

The rapidly increasing world population requires higher quantitative and qualitative agricultural productivity. Higher food crop yields have been achieved by improving soil productivity via the addition of fertilizers. This is one of the vital input materials for sustainable crop production. A major drawback of conventional fertilizers is their fast dissolution in soil relative to their absorptivity by plants. Consequently, water runoff results in the loss of fertilizer material and causes contamination of the surrounding environment.

Mineral fertilizers are some of the most important products of agricultural industry. While providing nutrients to crops, they increase their growth and at the same time they play an important role in regulating both pH and fertility of the soil. Production and consumption of mineral fertilizers have risen along with an increase in human population and a need for increased food production [51].

Controlled Release Fertilizers for Sustainable Agriculture
ISBN 978-0-12-819555-0
https://doi.org/10.1016/B978-0-12-819555-0.00004-2

In order to reduce the loss of the nutrients they supply and to improve the efficiency of chemical fertilizers, attention has been focused on the development of slow- and controlled-release fertilizers (SRFs and CRFs) in an attempt to minimize this difference between solubility and uptake [3]. Synthesizing process of SRFs and CRFs involves several physical methods such as dispersing ordinary fertilizers throughout a matrix and chemical methods such as encapsulating ordinary fertilizers within a larger compound. Both methods aim at slowing down the release of nutrients either by diffusion or dissolution (Wu et al., [52,62]).

2. Control- and slow-release fertilizers

SRFs and CRFs are described as materials that slowly release soluble nutrients over an extended period of time [Doxon., 1991., 93]. These fertilizers are easily contrasted against water-soluble fertilizers (WSFs), which are instantly soluble and leachable when applied to media [110]. Similarly, granular fertilizers (or dry, soluble fertilizer pellets) are also quickly soluble when exposed to moisture, can cause plant injury, are readily solubilize in media, and cause considerable leaching [11]. The likelihood of plant injury and nutrient leaching is reduced with C/SRFs [11]. SRFs and CRFs are used primarily in specialty markets such as nursery [woody-plant] production and compose only 0.15% of the current fertilizer market with a demand rising by approximately 5% annually [63,67].

SRFs action in the slow release of nutrients occurs primarily due to microbial activity and/or chemical hydrolysis [67]. In order for adequate release to occur, sufficient moisture and warm temperatures (generally above 20°C) are needed to initiate and boost microbial activity. SRFs are made of substances that are only slightly water soluble and require additional time for mineralization, thereby giving them their slow-release properties. SRFs may be organic or inorganic materials, and unlike CRFs they are not coated [93].

These fertilizers are often utilized in a nursery setting in order to reduce nutrient losses due to high irrigation or rainfall and a wide variability in temperature [11]. In contrast, such conditions are not likely to be encountered in a greenhouse setting, which is under constant environmental control [71,65]. In this context, it may be more difficult for greenhouse growers to recognize the potential advantages of fertilization with C/SRFs.

Researchers have developed types of CRF that consist of encapsulated solutions in a semipermeable membrane, surrounded by a hydrophobic polymeric or sulfur coating [8,67,93]. The coating on CRFs assists in limiting exposure of the core of the capsule to moisture, allowing for slow release of nutrients by diffusion [67,93]. Sulfur-based coatings consisting of sulfur-coated urea are relatively inexpensive.

The low solubility of CRFs allows for their release through imperfections and small pore openings based on the coating thickness [67]. However, the most advanced CRF products contain polymer coatings, which allow for even longer longevity and the ability to maintain constant release by adjusting not only the coating thickness but composition as well [67]. A thicker polymeric coating will result in a product with a more extended longevity. A linear release rate of nutrients often does not occur due to differences in temperature, moisture content, and variability in coating thickness and granule size [8,15].

A primary example of an organic slow-release fertilizer is compost. Inorganic examples include urea-based fertilizers (such as urea formaldehyde, isobutylidene diurea, and triazone), magnesium ammonium phosphates (MagAmps), and other materials which degrade biologically and are not readily soluble in water [31,93].

The relatively higher cost of C/SRFs may be the issue that causes many growers to be reluctant in purchasing and using CRFs in production [63,66]. However, each operation would benefit from conducting an in-house cost-benefit analysis. Due to a wide variability in formulations, crop demand, and necessitated application rates, this should be done on an operation-specific basis. Another concern growers may have is the current inability of CRFs, while they may constantly release nutrients, to match periods of high N demand [31,96]. On the other hand, CRFs of different longevities exhibit different release patterns during bedding plant production, thus highlighting the importance of selecting the proper product [4]. Additionally, environmental conditions may dictate a higher N release pattern or loss based on high rates of leaching, constant moisture or saturation, and high temperature and humidity [11,93]. The consequence of such release characteristics may equate to the loss of control of a fertility program for some growers.

2.1 Characterization of fertilizers with slow/control release of nutrients

SRFs and CRFs contribute to low-carbon economy and therefore are considered a significant advancement to the fertilizer industry in the 21st - century. The worldwide SRF market is believed to have to develop at a compound annual growth rate of 6.5% between 2014 and 2019. Based on geological areas, SRF market is classified into five key regions including Europe, Asia-Pacific, North America, Middle East, and Africa and Latin America [45,102].

[45] defined so-called "intelligent fertilizers" as fertilizers with a delayed release of mineral nutrients, according to the nutrient requirements of the plants. Examples of such materials are SRFs and CRFs which gradually release mineral components, simultaneously ensuring proper plant nutrition.

The economic, environmental, and physiological advantages related to the soil application SRFs and CRFs are discussed thoroughly elsewhere. Nutrient uptake by plants in their vegetation cycle has a sigmoidal character. An important application of SRFs/CRFs involves the release of their nutrients in a way to better fit the plants' requirements and ensures an improved effectiveness of fertilization while minimizing the loss between application and absorption. At the same time using SRFs/CRFs allows to reduce negative influence that fertilizers have on the environment largely due to high solubility of nitrogen compounds which are left unused.

Nowadays the most common classification is based on six types, as detailed in Table 4.1 [45].

2.2 Improved classification

Many previous references to the concept of SRFs date back to the 1960s [36,76,117]. More recently, there has been an almost exponential growth in publications on the benefits of SRFs to human society [16,28,33,55,64,88,92,104]. Despite the increase in publication numbers on SRFs, a systematic typology and comprehensive framework for SRFs remains not unified. Therefore, this part of our study aims at providing such an improved classification's framework, of which the main elements are presented in Fig. 4.1.

Numerous specialized researchers described two classification methods that depend on the principle of slow release and on the mode of dissolved release. In that sense, experience suggests the classification method based on the principle of slow-release control can cover the type of SRFs in an exhaustive manner. Consequently, the improved classification method is also based on the principle of slow release, which is divided into three types.

The traditional classification system is relatively backward, and the content begins to lag the new concepts. As a result, there are some defects

Table 4.1 Classification of slow release fertilizers.

Classification basis	Type 1	Type 2	Type 3	Type 4
Slow-/controlled-release principle [119]	Physical	Chemical	Physical and chemical [47]	×
Mode of dissolution release (Azeem et al. [20])	Organic compounds	Water-soluble, fertilizers with physical barrier	Inorganic low solubility compounds [90]	×
Class of nutrient elements [48]	Unit	Multivariate		×
Mode of nutrient release [21]	Diffusion	Erosive or chemical reaction	Swelling [95]	Osmosis
Production process and fertilizer properties Chen [7]	Coated (wrapped)	Inhibitor modified [85]	Chemical synthesis of slow soluble organic/inorganic nitrogen	Organic Matter [91] and matrix compound and adhesive
Chemical composition [57]	Polymerization or condensed SRF	Wrapped SRF	Mixed SRF	Adsorption SRF

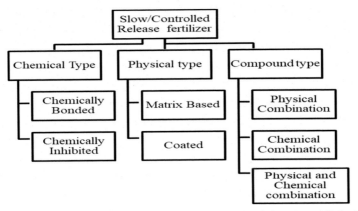

Figure 4.1 Classification of SRFs. *(Based on the principle of slow/ controlled-release.)*

in the classification of some new products. The proposed classification incorporates the physical type and chemical type of classification used by Ref. [119]. At the same time, it introduces the compound type for the first time, which satisfies the extensibility of SRFs classification. It solves the problem of the attribution of new compound fertilizer, which has shown high efficiency and slow–release performance in some literature [39,53].

In addition, Schiff base complex inhibitor is introduced into the chemical type to perfect the chemically inhibited fertilizers category and provide insight into the potential new SRF optimization design basis [45].

2.3 Types of slow/control-release fertilizers

[45] referred to the largest proportion of SRFs is consumed in nonagricultural markets (e.g., for lawn care, golf courses, landscaping), with an annual growth in demand of about 5%. The use of SRFs in agriculture slightly exceeds 10% of the total 10% [88].

The mainstream of fertilizers is mainly concentrated on physical and chemical types [60,84,103], and the use of compound type of SRFs has rarely been reported. In terms of the effect of use, the physical type of SRFs can overcome the shortcomings that rapid dissolution of fertilizer into the soil but cannot control the soil transformation behavior of the dissolved fertilizer [87]. The chemical type of SRFs can slow down the rate of enzymatic hydrolysis of fertilizer [17,24], but the delayed hydrolysis time is very short, even affected by soil type and crop variety. As for compound type of SRFs, it can more effectively control the process of dissolution and transformation of fertilizer in soil [121]. From one perspective, it can viably slow down the hydrolysis.

Then again, it can make the enzymatic hydrolysis process smoother, thus effectively avoiding the wonder of nutrient explosion. From the point of view of classification, the following is a detailed description of the utilization of SRFs belonging to their own sorts [45].

2.3.1 Physical type

Physical-type process belonging to SRFs can avoid the direct contact between fertilizer and soil by simple physical means like coating form or using matrix, in order to control the speed of the soil water to enter the fertilizer core and nutrient solution from the inside out, reaching the purpose of synchronization for the release rate with the nutritional demand of the plants [46]. Further, physical-type fertilizers are divided into coated fertilizers and matrix fertilizers. Physical type of fertilizers started early, are popular, but their slow-release properties are uneven.

[3] indicated that in order to produce fertilizers with slow release of nutrients by mechanochemistry applications as a physical method of preparation, the SRFs or CRFs are usually treated by the physical methods such as dispersion of ordinary fertilizer in the matrix or encapsulating familiar fertilizer. Nutrient release was retarded by diffusion and widely discussed in the literature [6, 98,123]. Many recent studies have reported the application of the mechanochemical synthesis approach to prepare complex compounds to be used as SRFs [6,98,123].

Most S/CRFs were prepared by using inorganic raw materials [61]; SRFs have been prepared during the intercalation of urea [NH_2CONH_3] into the kaolin structure by a mechanochemical process involving milling and compared against the aqueous suspension method. Also, [99] have successfully synthesized $KMgPO_4$ and NH_4MgPO_4 compounds as SRFs. [3] discussed the fine milling in applied mechanochemistry, and its application in agriculture for the production fertilizers. The experimental procedures for many prepared SRFs are illustrated in Fig. 4.2.

2.3.1.1 Coated fertilizers

In recent years, the rapid development of coated fertilizers, accounting for more than 95% of SRFs and CRFs. The process involves the spraying of a layer or several layers of inert material on the surface of the fertilizer particles to form a compact, low permeability film. The different properties of the membrane structure achieve different slow release effects. The most common coated slow-release fertilizers are sulfur- and polymer-coated urea [14]. The coating technology is considered as the most suitable method for

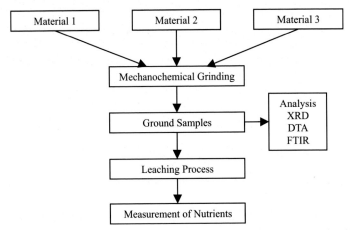

Figure 4.2 Flow chart of the experimental procedure of making SRF fertilizers.

consistent supply of the nitrogen to the plants and for reduction of loss and contamination effects [69]. [45] divided coated fertilizers into three types: inorganic coating, organic polymer coating, and multifunctional composite coating. Advantages and disadvantages of some commonly used coating materials are summarized in this chapter.

[113] studied and compared the nutrient release characteristics of sulfur-coated urea, humic acid-coated urea, and common urea in mild saline-alkaline soil and their effects on the absorption of nitrogen and yields of spring maize. The results indicated that it had the same trend in the mild saline-alkali soil within the water, but the nutrients release rate was 87.9% when the maize was harvested, which was higher than 59.3% in static water state. Compared with the two treatments, the treatments of sulfur-coated urea were better than the treatment of humic acid urea and common urea.

The increment rate of maize nitrogen absorption amount was 26.97% and 44.32%, respectively, the yield of corn increased by 8.73% and 22.43%. However, "locked-in effect" and "tailing effect" appeared in the later period. Although inorganic and natural polymers are still half the size of coated fertilizers, they are not always possible, still there are issues from the standpoint of obtaining predictable slow-release characteristics due to unwanted surface cracks, essential abrasion, and impact resistance and complexity of the fertilizer processing [45].

With the development of polymers, the later modified starch coating, various resin coating, and so on were found. A starch-based super-absorbent polymer (starch–SAP) coated slow–release fertilizer is developed [42, 82] and provided three sources of starch from maize, cassava, and potato. In the experiment, the fertilizer particles coated with starch–SAP displayed well slow-release behaviors.

[106] built up a biodegradable polymer coating to decide the applicability of various side effects and waste blends and to reduce coating costs. The composites consisting of polyvinyl alcohol, horn meal (slaughterhouse waste), rapeseed cake, crude glycerin (result), and phosphor gypsum were proposed for the encapsulation of mineral fertilizers. Moreover, it indicated extensive positive effect on the improvement of the roots of the tomato grows as well as enabled to enhance them with the nutrients helpful for the vegetation of plants.

2.3.1.2 Matrix-based fertilizers

Matrix-based fertilizers (MFs) is a novel type of SRFs. MFs essentially consist of matrix-based urea and matrix-based compound fertilizer, which are created based on comparative mechanisms and framework materials. Fertilizer nutrients are released by the dissolution of nutrients through functional materials and/or disintegration of their own materials. It tends to be separated into two sorts of nutrient adsorption and diffusion control. Although the release period of nutrient is shorter than that of coated fertilizer, it has the benefits of far reaching nutrient, low cost, and friendly environment [86, 124].

It is very suitable for field production and the requirement of pollution-free agriculture, also has a bright prospect in agriculture [73]. Three randomized block design of three treatments were conducted [120]: control test (CK, without urea application), common urea treatment (CU, 195 kg N/ha), and matrix-based urea treatment (MU, 195 kg N/ha). Additionally, two laboratory tests were conducted to assess N leaching and ammonia emission from matrix-based urea.

The results showed the MU group, which was mixed with bentonite, organic polymer, and urea, was superior to the normal urea treatment group. Grain yields with MU were 6.3% and 14.7% greater than those with CU, respectively; agronomic efficiency and apparent recovery efficiency were greater with MU than with CU. The glass–matrix based fertilizer (GMF, a byproduct from ceramic industries) was applied [105] on "Tarocco" orange trees. The GMF mixture showed to be able to supply adequately micronutrients (particularly Fe) on a long term, reducing the chlorosis symptoms, increasing the leaf SPAD index, and decreasing

Fe index. The GMF not only reduced the use of chemicals (such as Fe-chelate), but also reused industrial wastes and organic residues which give an "adding value" to these novel organomineral formulates. Although little work has been done on using this general nanotechnology-based principle for delivering nutrients to crops [50], focused on the synthesis of environmentally benign nanoparticles carrying urea as the crop nutrient that can be released in the soil.

2.3.2 Chemical type

In conventional fertilizing (e.g., with urea) nutrients release lasts 30—60 days, which given a 100—120-days-long crops growth cycle means that a fertilizer must be applied 2 or 3 times. SRF/CRFs release their nutrients slowly and gradually during all vegetation season and consequently need to be applied once only, which greatly reduces both time and energy consumption [71].

A superior and more efficient utilization of nutrients can lead both to decrease of waste material produced by the fertilizers industry and to a decrease in gaseous petrol and different assets utilization.

According to the Association of American Plant Food Control Officials (AAPFCO) [5], SRFs are chemically or biologically decomposed martials with a high molecular weight, complex structure, and little dissolvability in water, CRFs are materials of which the release of mineral components takes place through a polymer layer or a film [5].

Chemical type of SRFs can be divided into two categories: chemically bonded fertilizers and chemically inhibited fertilizers. The effect of this kind of SRFs is better, but the production cost is relatively high [45].

2.3.2.1 Chemically bonded fertilizers

Chemically bonded fertilizers are those that require the application of fertilizers to be combined with one or more chemical materials by covalent or ionic bonding to produce a slightly soluble or water–insoluble compound. It releases nutrients slowly through the direct action of crop root system and microbial degradation. The rate of nutrient release was mainly determined by the size of the particles, the water content in soil, temperature, and pH [119].

Moreover, It is also affected by the chemical structure of the material, the strength of covalent or ionic bond, the degree of polymerization and environmental degradation. The common chemical materials are urea formaldehyde, isobutylidenediurea, polyphosphate, long-lasting silicate potassium fertilizer, etc. [45].

[115] compared the effects of five controlled-release nitrogen fertilizers, including isobutyldiurea (IBDU), sulfur-coated urea (SCU), urea formaldehyde (UF), methylene urea (MU), and polyolefin-coated urea (ESN). Experiments showed that the five kinds of SRF/CRFs increased the leaf chlorophyll content, net photosynthetic rate, transpiration rate, nitrogen content, and root activity at different degrees and found that the order of the studied fertilizers' effects from high to low is MU > IBDU > UF > ESN > SCU. Consequently, the chemical slow/controlled fertilizers are better than physical type.

[41] used a 16S rRNA gene clone library, the study found that the urea-formaldehyde fertilizer markedly increased bacterial diversity in onion bulbs and main roots of sugar beet. The results of principal coordinates analysis revealed that the community structures in both plants shifted unidirectional in response to the UF fertilizer, which also indicated the UF fertilizer may be used as a driving force to manipulate a bacterial community in plants.

Guo et al. [32] prepared successfully a new type of slow-release nitrogen (N) and phosphorus (P) compound fertilizer with superabsorbent and moisture preservation qualities by using carboxyl methyl starch, acrylic acid, ammonia, urea, diammonium phosphate, and so on. The product was characterized by Fourier-transform infrared spectroscopy (FTIR), inductively coupled plasma (ICP), and element analysis, and the results showed that it contained 22.6% nitrogen element and 7.2% phosphor (showed by P_2O_5) element, and the N and P in it were in the form of urea and $(NH_4)_2HPO_4$. The water-holding capacity results confirmed that the largest water-holding ratio was 12% higher than that without it, and the water retention experiment still had an excellent moisture preservation capacity in soil. Consequently, the slow-release ratio of the effective nutrient in it was not above 75% on the 30th day and the product could not only efficiently improve the utilization efficiency of fertilizers but also the water resource.

2.3.2.2 Chemically inhibited fertilizers

The chemically inhibited fertilizers can release nitrogen slowly through the addition of inhibitors; commonly used are urease inhibitors and nitrification inhibitors, which slow down the hydrolysis of urea, and the latter retard the nitrification of ammonium. Common inhibitors are hydroquinone, acetohydroxamic acid, pyridine, dicyandiamide, etc. Although there are many related inhibitors reported at present, but most of the inhibitors (such as metal salts and organic small molecules) have the negative environmental effects such as high toxicity, easy loss, short efficacy, and so on [38, 45].

A field experiment with lettuce was conducted in the Savanna of Bogotá [40]. It showed that 3,4-dimethylpyrazole phosphate [DMPP] does not affect greenhouse gases emissions in lettuce crop. Additionally, a reduction of 20 kg N/ha using DMPP was able to keep the yield mean while improving the quality of the crop. The nitrification inhibitors' (NIs) effects on maize yields were investigated in a loamy-sand soil in Thailand [79]. Results indicated the applied NIs increased the maize yield by 13%–20% and 17%–24% for maize grain and whole plant.

As urease is the specific enzyme which leads to the decomposition of urea, and in view of the disadvantages of the existing three types of urease inhibitors, the research and development of the fourth type of urease inhibitor in recent years has become a hot topic at home and abroad. This is because this new kind of urease inhibitor mostly uses Schiff base organic molecules, with important pharmacological and biological activities, and inorganic metal salts, required for essential plant growth as molecular building blocks.

Through a variety of coordination ligand modifications, flexible regulation, and activity screening, the fourth Schiff base complex inhibitor with double active sites was constructed [35,78,97].

[112] prepared a new copper (II) complex which is an azide-bridged poly nuclear with Schiff base 2-bromo-4-chloro-6-([2-dimethylamino ethylimino]methyl) phenol. As shown by data, the copper (II) complex exhibits strong urease inhibitory property, with the IC_{50} value of 13.0 ± 0.4 being much lower than that of acetohydroxamic acid coassayed as a standard reference against urease.

Polymer-coated controlled-release fertilizers(PCRF) are the newest and most technically sophisticated fertilizers being utilized in plant generation and production and comprise of a core of soluble nutrients encompassed by a polymer coating, and nutrient release is precisely controlled by the chemical composition and thickness of the polymer coating. In contrast with the past classifications that ony supply nitrogen, PCRFs supply all three "fertilizer elements" N, P, and K, and many formulations include calcium, magnesium, sulfur, and micronutrients. Slow-release fertilizer can also be achieved with a composition based on urea-modified hydroxyapatite nanoparticles encapsulated wood [49].

These studies recommended that there is a distinct advantage using polymer coatings contrasted to other coatings. Thus, it is considered to carry out the slow release of potash fertilizer through a coating of polyacrylamide polymer. Polyacrylamide is chosen as the polymer since it reduces soil disintegration [74].

2.3.3 Compound type

It belongs to the compound type of SRFs, which is a sort of fertilizer prepared by a comprehensive combination of physical or/and chemical techniques. The fertilizer composition contains at least two of the previously mentioned types of SRFs. It can be divided into three classes:

a. Physical and chemical combination of fertilizer (comprehensive utilization of physical and chemical methods).

b. Physical combination of fertilizer [coated with matrix method], i.e., encapsulated SRF. Zhang et al. studied SRFs encapsulated by rapheme oxide films, and the results showed that new coating technology could hold great promise for the development of environmentally benign CRFs for crop production [72, 122].

c. Chemical combination of fertilizer (chemical combination with inhibitor method). To a certain extent, this kind of fertilizer can completely control the process of nutrient dissolution and transformation in soil, which has great development potential and broad prospect.

Among them, there are few reports on physical combination of fertilizer, but considering the rapid rate of fertilizer replacement, the corresponding categories are reserved. As for chemical combination of fertilizer, it appears more in the form of a patent. Physical and chemical combination fertilizer has been explored in some literature [2,13,25,54], and has been proved to have a synergistic effect.

[26] invented a composition comprising a combination of substantially solid pieces of N-(n-butyl) thiophosphoric triamide (NBPT) and urea-formaldehyde polymer (UFP) particles, in which the NBPT and the UFP particles are substantially adhered together. The formulation of the present invention significantly improves the flow of the mixture comprising the urease inhibitor, thereby reducing dust, increasing bulk density, and providing a more uniform formulated product. The fertilizer composition of this invention is capable of supplying the nitrogen nutrient with greater efficiency than any previously known fertilizer composition [118].

[1] was conducted to determine ammonia loss from surface-broadcast urea, urea treated with three rates of NBPT, and experimental zinc sulfate ($ZnSO_4$)- coated urea (ZSCU) fertilizers with or without urease inhibitors. Laboratory and field experimental results demonstrated that cumulative ammonia loss from urea was greater than all rates of NBPT-treated urea (5.57%–13.35%) and ZSCU fertilizer containing urease inhibitors (2.63%–11.50%) across soil types. The results also indicated

that the ZSCU fertilizers have the potential to reduce ammonia losses as compared to urea; however, ZSCU fertilizers with B compounds (H_3BO_3) and NBPT were the most effective.

[83] studied a novel slow-release NPK fertilizer encapsulated by superabsorbent nanocomposite involving the mineral clay montmorillonite. The results indicated that the presence of the montmorillonite caused the system to liberate the nutrient in a more controlled manner than that with the neat superabsorbent. The good SRF property as well as good water retention capacity showed that this formulation is potentially viable for application in agriculture as a fertilizer's carrier vehicle [52, 83].

3. Advantages of using CRFs and SRFs

The use of C/SRFs in crop production has demonstrated much success. There are many advantages to using C/SRFs in bedding plant production and other agricultural crops including:

(1) Nutrients are better utilized when slowly released throughout a season rather than applied in "bursts" or instantly soluble applications such as is the case in WSF (water-soluble fertilizer) application, thus increasing nutrient use efficiency and perhaps more closely synchronizing release rates with plant demand [22,65–67].

(2) The quantity of fertilizer used is also reduced, leading to less of a risk for plant injury through highly soluble salt levels [11,22].

(3) Nutrient leaching is greatly reduced when using C/SRFs as compared to water-soluble fertilizers (WSFs) [9,22,110].

(4) One fertilizer application may meet the seasonal need of the crop, reducing the need for additional labor [66].

(5) Crops such as poinsettia, begonia, and New Guinea impatiens produced using CRFs are shown to be of comparable or better quality than those produced with WSFs [9,81].

(6) Enhanced nutrient-use efficiency and decreased nutrient losses. The application of CRFs and SRFs can decrease fertilizer use by 20–30% of the recommended rate of a traditional fertilizer while obtaining the same yield [107].

(7) Reduced application and labor costs. For example, in current practices, commercial potato producers use three to four applications of nitrogen fertilizers for NE Florida and two applications for SW Florida (personal communication with local potato producers). Eliminating extra applications of fertilizer saves the farmer between $5 and $7/acre broadcasting expense [56].

(8) Minimization of fertilizer-associated risks such as leaf burning, water contamination, and eutrophication (a process where water bodies receive excess nutrients). The slow rates of nutrient release can keep available nutrient concentrations in soil solution at a lower level, reducing runoff and leaching losses [94].

(9) Better understanding of nutrient release rate and duration (CRFs only, because they are less sensitive to soil and climate conditions) [107]. Knowing when to apply fertilizer and in what quantities saves money, reduces fertilizer-associated risks to crops and the environment, and improves nutrient management programs.

(10) Lowered soil pH in alkaline soils for better bioavailability of some nutrients. Applying sulfur-coated urea will probably increase soil acidity because both sulfur and urea contribute to increasing the acidity (lowering soil pH) of the soil. Consequently, phosphorus or iron may be more bioavailable and benefit some crops like blueberry, potato, and sweet potato [58]. In addition, sulfur is an essential nutrient for all crops.

(11) Reduced production costs if there is an abundant supply of SRF sources like manures nearby.

4. Conclusions

The development and application of SRFs/CRFs have become a hot topic in the field of plant nutrition and fertilizer science worldwide. SRFs are used in a limited fashion in practice due to its high cost and difficult degradation of materials, mainly for high-value crops. This chapter points to the need for classification studies in order to provide evidence on which to base the selection and further develop innovative and enhanced efficiency SRFs.

At present, SRFs are divided into two main types: physical and chemical. Upon application, the physical type of SRFs have more varieties compared to the chemical type. From the cost point of view, the price of chemical SRFs is relatively higher. But when it comes to the production process, the physical type of SRFs are much more complex. Although a compound (mixed) type of SRF is not the mainstream right now, demand for compound type of fertilizer will experience continued growth as they prove to be efficient. Therefore, it is necessary to develop compound type of SRFs which is produced with ecofriendly and renewable materials with ample supply with somehow acceptable cost.

Through physical, chemical, and biochemical methods, the compound type of SRFs which have obvious advantages are obtained by combining some Schiff base complex inhibitors within a certain coating. Future research involving compound type of SRFs is likely going to improve yields and nitrogen fertilizer recovery. It is of great scientific significance and application value for maintaining food security, developing highly efficient ecological agriculture, reducing rural nonpoint pollution, and promoting sustainable agriculture development.

References

[1] N. Adotey, M. Kongchum, J. Li, G.B. Whitehurst, E. Sucre, D.L. Harrell, Ammonia volatilization of zinc sulfate- coated and NBPT-treated urea fertilizers, Agron. J. 109 (6) (2017) 2918−2926, https://doi.org/10.2134/agronj2017.03.0153.

[2] A.E. Al-Rawajfeh, M.R. Al-rbaihat, E. AlShamaileh, Clean and efficient synthesis using mechanochemistry: preparation of kaolinite KH_2PO_4 and kaolinite−[$NH_4]_2HPO_4$ complexes as slow released fertilizer, J. Ind. Eng. Chem. 73 (2019) 336−343.

[3] E. AlShamaileh, M.R. Al-rbaihat, Al-Rawajfeh, A. E, Mechanochemical synthesis of slow-release fertilizers: a review, Open Agric. J. 12 (2017) 11−19, https://doi.org/10.2174/1874331501812010011.

[4] G.A. Andiru, Effects of Controlled-Release Fertilizer on Nutrient Leaching and Garden Performance of Impatiens Walleriana [Hook F. "Extreme Scarlet"], Master"s Thesis. Dept of Hort. and Crop Sci. Ohio St. Univ, 2010.

[5] Association of American, Plant Food Control Officials [AAPFCO], Official Publication No. 50, T-29. West Lafayette, IN, USA, 1997 (AAPFCO).

[6] P. Baláž, in: Mechanochemistry in Nanoscience and Minerals Engineering, Springer Verlag, BerlinHeidelberg, 2008.

[7] J.P. Chen, Preparation and properties of clay/polyacrylic acid (potassium polyacrylate) used for controlled or slow-release fertilizer, [M.D. Dissertation] Heifei University of Technology (2012) 1−52.

[8] B.A. Birrenkott, J.L. Craig, G.R. McVey, A leach collection system to track the release of nitrogen from controlled-release fertilizers in container ornamentals, HortScience 40 (2005) 1887−1891.

[9] E.K. Blythe, J.L. Mayfield, B.C. Wilson, E.L. Vinson, J.L. Sibley, Comparison of three controlled-release nitrogen fertilizers in greenhouse crop production, Plant Nut. 25 (2002) 1049−1061.

[10] Deleted in review.

[11] T.K. Broschat, Nitrate, phosphate, and potassium leaching from container-grown plants fertilized by several methods, HortScience 30 (1995) 74−77.

[12] Deleted in review.

[13] X. Chen, Z.W. Liao, X.Y. Mao, D.H. Wang, Technology and effect of controlled-release fertilizer through physical-biochemical composite approaches, J. Huazhong Agric. University 28 (3) (2009) 300−305 (Chinese).

[14] Q. Cheng, Effect of Real and Simulated Traffic on Coated Fertilizer Prill Integrity and Nitrogen Release, Auburn University, Alabama, 2017, pp. 1−71.

[15] S.H. Chien, L.I. Prochnow, H. Cantarella, Recent developments of fertilizer production and use to improve nutrient efficiency and minimize environmental impacts, Adv. Agron. 102 (2009) 267–322.

[16] P. Chawakitchareon, R. Anuwattana, J. Buates, Production of slow release fertilizer from waste materials, Advanced Materials Springer International Publishing 84 (4) (2016) 534–540.

[17] A.T.M.A. Choudhury, I.R. Kennedy, Nitrogen fertilizer losses from rice soils and control of environmental pollution problems, Commun. Soil Sci. Plant Anal. 36 (11–12) (2005) 1625–1639, https://doi.org/10.1081/CSS-200059104.

[18] Deleted in review.

[19] Deleted in review.

[20] B. Azeem, K. Kushaari, Z.B. Man, A. Basit, T.H. Thanh, Review on materials & methods to produce controlled release coated urea fertilizer, Journal of Controlled Release 181 (2014) 11–21.

[21] T. Emilsson, J.C. Berndtsson, J.E. Mattsson, K. Rolf, Effect of using conventional and controlled release fertiliser on nutrient runoff from various vegetated roof systems, Ecol. Eng. 29 (3) (2007) 260–271, https://doi.org/10.1016/j.ecoleng.2006.01.001.

[22] M.E. Engelsjord, O. Fostad, B.R. Singh, Effects of temperature on nutrient release from slow-release fertilizers, Nutrient Cycl. Agroecosyst. 46 (1997) 179–187.

[23] Deleted in review.

[24] L.D.A. Faria, C.A.C.D. Nascimento, G.R. Bardella, T.A.D. Moura, F.L. Mendes, G.C. Vitti, NH_3 volatilization from urea-NBPT in eucalyptus, Commun. Soil Sci. Plant Anal. 47 (6) (2016) 769–774, https://doi.org/10.1080/00103624.2016.1146892.

[25] W.H. Frame, M.M. Alley, G.B. Whitehurst, B.M. Whitehurst, R. Campbell, In vitro evaluation of coatings to control ammonia volatilization from surface-applied urea, Agron. J. 104 (5) (2012) 1201–1207, https://doi.org/10.2134/agronj2012.0009.

[26] K.D. Gabrielson, S.L. Wertz, D.R. Bobeck, A.R. Sutton, Compositions of Urea Formaldehyde Particles and Methods of Making ther Eof, 2017. US009682894B2.

[27] Deleted in review.

[28] A.S. Giroto, G.G. Guimarães, M. Foschini, C. Ribeiro, Role of slow-release nanocomposite fertilizers on nitrogen and phosphate availability in soil, Sci. Rep. 7 (46032) (2017) 1–11, https://doi.org/10.1038/s41598-016-0028-x.

[29] Deleted in review.

[30] M.V. Gold, Sustainable Agriculture: Definitions and Terms, USDA, 2007. http://www.nal.usda.gov/afsic/pubs/terms/srb9902.shtml.

[31] E.A. Guertal, Slow-release nitrogen fertilizers in vegetable production: a review, HortTechnology 19 (2009) 16–19.

[32] M. Guo, M. Liu, Z. Hu, F. Zhan, L. Wu, Preparation and properties of a slow release NP compound fertilizer with superabsorbent and moisture preservation, J. Appl. Polym. Sci. 96 (2005) 2132–2138.

[33] C. Guo, P. Li, J. Lu, T. Ren, R. Cong, X. Li, Application of controlled-release urea in rice: reducing environmental risk while increasing grain yield and improving nitrogen use efficiency, Commun. Soil Sci. Plant Anal. 47 (9) (2016) 1176–1183, https://doi.org/10.1080/00103624.2016.1166235.

[34] Deleted in review.

[35] L. Habala, S. Varényi, A. Bilková, P. Herich, J. Valentová, J. Kožíšek, F. Devínsky, Antimicrobial activity and urease inhibition of Schiff bases derived from isoniazid and fluorinated benzaldehydes and of their copper [II] complexes, Molecules 21 (2016) 1742, https://doi.org/10.3390/molecules21121742.

[36] L.I. Hansen, Slow Release Fertilizer Granule Having a Plurality of Urethane Resin Coatings US3264089A, 1966.

[37] Deleted in review.

[38] R.D. Hauck, in: O.P. Engelstad (Ed.), Slow-release and Bio Inhibitor-Amended Nitrogen Fertilizers, Fertilizer Technology and Use, vol. 3, 1985, pp. 293—322.

[39] J. Hou, Y. Dong, G. Li, Z. Fan, S. Liu, N-release Characteristics of coated compound fertilizers added by nitrification inhibitor and their effects on Chinese cabbage, J. Soil Water Conserv. 25 (2011) 250—253 (Chinese).

[40] X.S. Huérfano, M.M. Menéndez, M.B. Bolaños-Benavides, J.M. González-Moro, C. Estavillo, González- Murua, The nitrification inhibitor 3, 4-dimethyl-pyrazole phosphate decreases leaf nitrate content in lettuce while maintaining yield and N_2O emissions in the Savanna of Bogotá, Plant Soil Environ. 62 (2016) 533—539, https://doi.org/10.17221/105/2016-PSE.

[41] S. Ikeda, K. Suzuki, M. Kawahara, M. Noshiro, N. Takahashi, An assessment of urea-formaldehyde fertilizer on the diversity of bacterial communities in onion and sugar beet, Microb. Environ. 29 (2014) 231—234, https://doi.org/10.1264/jsme2.ME13157.

[42] S. Irfan, R. Razali, K.Z. Kushaari, N. Mansor, Reaction-Multi diffusion model for nutrient release and autocatalytic degradation of PLA-coated controlled-release fertilizer, Polymers 9 (2017) 111—123, https://doi.org/10.3390/olym9030111.

[43] Deleted in review.

[44] Deleted in review.

[45] J. Fu, C. Wang, X. Chen, Z. Huang, D. Chen, Classification research and types of slow controlled release fertilizers [SRFs] used - a review, Commun. Soil Sci. Plant Anal. (2018), https://doi.org/10.1080/00103624.2018.1499757.

[46] J.Y. Jiang, Y.L. Fu, Prospect for physical type slow/controlled release fertilizers, World J. For. 2 (2013) 35—39, https://doi.org/10.12677/WJF.2013.24007 (Chinese).

[47] Y.Y. Jin, R.F. Sun, Y. Su, J. Xu, S. Zhao, X.B. Cheng, Research status of coated slow- and controlled-release fertilizers at home and abroadv[J], Vegetables 9 (2016) 40—44.

[48] L.S. Kasmaei, M. Fekri, Application of Cu fertilizer on Cu recovery and desorption kinetics in two calcareous soils, Environ. Earth Sci. 67 (2012) 2121—2127, https://doi.org/10.1007/s12665-012-1652-9.

[49] N. Kottegoda, I. Munaweera, N. Madusanka, V.K.A. Green, Slow-release fertilizer composition based on urea-modified hydroxyapatite nanoparticles encapsulated wood, Curr. Sci. 101 (2) (2011) 1—7.

[50] N. Kottegoda, C. Sandaruwan, G. Priyadarshana, A. Siriwardhana, U.A. Rathnayake, D.M. Berugoda Arachchige, A.R. Kumarasinghe, D. Dahanayake, V. Karunaratne, G.A. Amaratunga, Urea-hydroxyapatite nanohybrids for slow release of nitrogen, ACS Nano 11 (2) (2017) 1214—1221, https://doi.org/10.1021/acsnano.6b07781.

[51] L.H. Xie, M.Z. Liu, B.L. Ni, X. Zhang, Y.F. Wang, Slow-release nitrogen and boronfertilizers from a functional superabsorbent formulation based on wheat straw and attapulgite, Chem. Eng. J. 167 (2011) 342—348.

[52] L. Wu, M.Z. Liu, R. Liang, Preparation and properties of a double-coated slow release NPK compound fertilizers with superabsorbent and water-retention, Bio resource, Technol. 99 (2008) 547—554.

[53] D.P. Li, Z.J. Wu, C.H. Liang, Nitrogen release characteristics of coated urea amended with biological inhibitors in meadow brown soil, Plant Nutr. Fert. Sci. 16 (1) (2010) 214—218.

[54] D.P. Li, Z.J. Wu, C.H. Liang, Traits of urea nitrogen release from starch acetate coated urea and amended with biological inhibitors in meadow brown soil, Chin. J. Soil Sci. 42 (2011) 1376—1381.

[55] Y.M. Li, Y.X. Sun, S.Q. Liao, G.Y. Zou, T.K. Zhao, Y.H. Chen, J.G. Yang, L. Zhang, Effects of two slow- release nitrogen fertilizers and irrigation on yield, quality, and water-fertilizer productivity of greenhouse tomato, Agric. Water Manag. 186 (2017) 139—146, https://doi.org/10.1016/j.agwat.2017.02.006.

[56] G.D. Liu, H.S. Eric, Y. Li, M.H. Chad, M. Warren, S. Lands, Controlled-Release Fertilizer Opportunities and Costs for Potato Production in Florida, Publication no.HS941, UF/IFAS Extension, 2011.

[57] Y.Y. Liu, The Preparation of New Coated Slow-Release Fertilizer and Properties Study, Nanjing University of Science & Technology, 2012, pp. 1–55 [M.D.Dissertation].

[58] G. Liu, E. Hanlon, Soil pH Range for Optimum Commercial Vegetable Production, Publication #HS1207, one of a series of the Horticultural Sciences Department, UF/IFAS Extension, 2012.

[59] Deleted in review.

[60] S.S. Malhi, E. Oliver, G. Mayerle, G. Kruger, K.S. Gill, Improving effectiveness of seed row-placed urea with urease inhibitor and polymer coating for durum wheat and canola, Commun. Soil Sci. Plant Anal. 34 (11–12) (2003) 1709–1727, https://doi.org/10.1081/CSS-120021307.

[61] E. Makó, J. Kristóf, E. Horváth, V. Vágvölgyi, Kaolinite–urea complexes obtained by mechanochemical and aqueous suspension techniques—a comparative study, J. Colloid Interface Sci. 330 (2009) 367–373.

[62] M. Teodorescu, A. Lungu, P.O. Stanescu, C. Neamtu, Preparation and properties of novel slow-release NPK agrochemical formulations based on poly [acrylic acid] hydrogels and liquid fertilizers, Ind. Eng. Chem. Res. 48 (2009) 6527–6534.

[63] L.C. Medina, J.B. Sartain, T.A. Obreza, Estimation of release properties of slow-release fertilizer materials, HortTechnology 19 (2009) 13–15.

[64] B.P. Meena, K. Ramesh, S. Neenu, P. Jha, I. Rashmi, Controlled Release Fertilizers for Improving Nitrogen Use Efficiency, New India Publishing Agency, New Delhi, 2017, pp. 59–79.

[65] D.J. Merhaut, E.K. Blythe, J.P. Newman, J.P. Albano, Nutrient release from controlled-release fertilizers in an acid substrate in a greenhouse environment: leachate electrical conductivity, pH, and nitrogen, phosphorous, and potassium concentrations, HortScience 41 (2006) 780–787.

[66] R.L. Mikkelsen, T.W. Bruulsema, Fertilizer use for horticultural crops in the during, U.S. During the 20th century, HortTechnology 15 (2005) 24–30.

[67] K.T. Morgan, K.E. Cushman, S. Sato, Release mechanisms for slow- and controlled-release fertilizers and strategies for their use in vegetable production, HortTechnology 19 (2009) 10–12.

[68] Deleted in review.

[69] M.Y. Naz, S.A. Sulaiman, Slow release coating remedy for nitrogen loss from conventional urea: a review, J. Contr. Release 225 (2016) 109–120, https://doi.org/10.1016/j.jconrel.2016.01.037.

[70] Deleted in review.

[71] P.V. Nelson, C. Niedziela, D. Pitchay, Efficacy of soybean-base liquid fertilizer for greenhouse crops, Plant Nutr. 33 (2010) 351–361.

[72] B. Ni, M. Liu, S. Lü, L. Xie, Y. Wang, Environmentally friendly slow-release nitrogen fertilizer, J. Agric. Food Chem. 59 (2011) 10169–10175, https://doi.org/10.1021/jf202131z.

[73] X.Y. Ni, Y.J. Wu, Z.Y. Wu, L. Wu, G.N. Qiu, L.X. Yu, A novel slow-release urea fertiliser: physical and chemical analysis of its structure and study of its release mechanism, Biosyst. Eng. 115 (2013) 274–282, https://doi.org/10.1016/j.biosystemseng.2013.04.001.

[74] K.N. Nwankwo, Polyacrylamide as a Soil Stabilizer for Erosion of Soil, Report WI06-98, 2001.

[75] Deleted in review.

[76] J.J. Oertli, O.R. Lunt, Controlled release of fertilizer minerals by incapsulating membranes: I. Factors influencing the rate of release, Soil Sci. Soc. Am. J. 26 (1962) 579−583, 10.2136/sssaj1962.03615995002600060019x.

[77] Deleted in review.

[78] L. Pan, C.F. Wang, K. Yan, K.D. Zhao, G.H. Sheng, H.L. Zhu, X.L. Zhao, D. Qu, F. Niu, Z.L. You, Synthesis, structures and helicobacter pylori urease inhibitory activity of copper[II] complexes with tridentate aroylhydrazone ligands, J. Inorg. Biochem. 159 (2016) 22−28.

[79] P. Pengthamkeerati, A. Modtad, Nitrification inhibitor effects on nitrous oxide emission, nitrogen transformation, and maize [Zea mays L.] yield in loamy sand soil in Thailand, Commun. Soil Sci. Plant Anal. 47 (7) (2016) 875−887, https://doi.org/10.1080/00103624.2016.1159314.

[80] Deleted in review.

[81] D.L. Richards, D.W. Reed, New Guinea impatiens growth response and nutrient release from controlled-release fertilizer in a recirculating subirrigation and top-watering system, HortScience 39 (2004) 280−286.

[82] D. Qiao, H. Liu, L. Yu, X. Bao, G.P. Simon, E. Petinakis, L. Chen, Preparation and characterization of slow- release fertilizer encapsulated by starch-based superabsorbent polymer, Carbohydr. Polym. 147 (2016) 146−154, https://doi.org/10.1016/j.carbpol.2016.04.010.

[83] A. Rashidzadeh, A. Olad, Slow-released NPK fertilizers encapsulated by NaAlg-g-poly[AA-co-AAm]/MMT superabsorbent nanocomposite, Carbohydr. Polym. 114 (2014) 269−278.

[84] H.M.K. Ross, A.B. Middleton, P.G. Pfiffner, E. Bremer, Evaluation of polymer-coated urea and urease inhibitor for winter wheat in southern Alberta, Agron. J. 102 (4) (2010) 1210−1216, https://doi.org/10.2134/agronj2009.0194.

[85] H. Sabahi, A.H. Rezayan, S. Sadeghi, S. Jamehdor, Study the N turnover of legume seed meals for designing a slow-release nitrogen fertilizer, Commun. Soil Sci. Plant Anal. 45 (10) (2014) 1325−1335, https://doi.org/10.1080/00103624.2013.875198.

[86] B.K. Saha, M.T. Rose, V. Wong, T.R. Cavagnaro, A.F. Patti, Hybrid brown coal-urea fertiliser reduces nitrogen loss compared to urea alone, Sci. Total Environ. (2017) 601−602, https://doi.org/10.1016/j.scitotenv.2017.05.270, 1496−504.

[87] K.L. Sahrawat, Forms, Properties and dissolution of controlled-release nitrogenous fertilisers, Fertil. News 40 (1995) 41−46, https://doi.org/10.1007/BF00749519.

[88] M.H. Sazzad, M.T. Islam, F. Chowdhury, A Review & Outlook of Slow-Release Fertilizer, LAP LAMBERT Academic Publishing, Germany, 2013, pp. 1−156, 978-3659415074.

[89] Deleted in review.

[90] S.I. Sempeho, H.T. Kim, E. Mubofu, A. Hilonga, Meticulous overview on the controlled release fertilizers, Adv. Chem. 1 (2014) 1−16, https://doi.org/10.1155/2014/363071.

[91] P. Sharrock, M. Fiallo, A. Nzihou, M. Chkir, Hazardous animal waste carcasses transformation into slow release fertilizers, J. Hazard Mater. 167 (1) (2009) 119−123, https://doi.org/10.1016/j.jhazmat.2008.12.090.

[92] A. Shaviv, Advances in controlled-release fertilizers, Adv. Agron. 71 (1) (2001) 1−49.

[93] A. Shaviv, R.L. Mikkelsen, Controlled-release fertilizers to increase efficiency of nutrient use and minimize environmental degradation-A review, Fert. Res. 35 (1993) 1−12, https://doi.org/10.1007/BF00750215.

[94] A. Shaviv, Controlled Release Fertilizers. IFA International Workshop on Enhanced-Efficiency Fertilizers, Frankfurt, International Fertilizer Industry Association Paris, France, 2005.

[95] Y.Z. Shen, C.W. Du, J.M. Zhou, F. Ma, Modeling nutrient release from swelling polymer-coated urea, Appl. Eng. Agric. 31 (2) (2015) 247−254.

[96] E.H. Simonne, C. Hutchinson, Controlled-release fertilizers for vegetable production in the era of best management practices: teaching new tricks to an old dog, Hort-Technology 15 (2005) 36—46.

[97] B.B. Sokmen, N. Gumrukcuoglu, S. Ugras, H. Sahin, Y. Sagkal, H.I. Ugras, Synthesis, antibacterial, antiurease, and antioxidant activities of some new 1,2,4-triazole Schiff base and amine derivatives, Appl. Biochem. Biotechnol. 175 (2015) 705—714, https://doi.org/10.1007/s12010-014-1307-2.

[98] Q.W. Solihin, W. Zhang, F. Tongamp, Saito, Mechanochemical synthesis of kaoline— KH_2PO_4 and kaoline—$NH_4H_2PO_4$ complexes for application as slow release fertilizers, Powder Technol. 212 (2011) 354—358.

[99] Q.W. Solihin, W. Zhang, F. Tongamp, Saito, Mechanochemical route for synthesizing $KMgPO_4$ and NH_4MgPO_4 for application as slow-release fertilizers, Ind. Eng. Chem. Res. 49 (2010) 2213—2216.

[100] Deleted in review.

[101] Deleted in review.

[102] Strategic Report, Controlled Release Fertilizers Market: Global Market Estimation, Dynamics, Regional Share, Trends, Competitor Analysis 2012 to 2016 and Forecast 2017 to 2023. Precision Business Insights [PBI], 2017, pp. 1—209. London, England. Website: www.precisionbusinessinsights.com.

[103] Z. Tian, J.J. Wang, S. Liu, Z. Zhang, S.K. Dodla, G. Myers, Application effects of coated urea and urease and nitrification inhibitors on ammonia and greenhouse gas emissions from a subtropical cotton field of the Mississippi delta region, Sci. Total Environ. 533 (2015) 329—338, https://doi.org/10.1016/j.scitotenv.2015.06.147.

[104] Y.P. Timilsena, R. Adhikari, P. Casey, T. Muster, H. Gill, B. Adhikari, Enhanced efficiency fertilisers: a review of formulation and nutrient release patterns, J. Sci. Food Agric. 95 (6) (2015) 1131—1142, https://doi.org/10.1002/jsfa.6812.

[105] B. Torrisi, A. Trinchera, E. Rea, M. Allegra, G. Roccuzzo, F. Intrigliolo, Effects of organo-mineral glass- matrix based fertilizers on citrus Fe chlorosis, Eur. J. Agron. 44 (2013) 32—37, https://doi.org/10.1016/j.eja.2012.07.007.

[106] J. Treinyte, V. Grazuleviciene, R. Paleckiene, J. Ostrauskite, L. Cesoniene, Biodegradable Polymer Composites as Coating Materials for Granular, 2017.

[107] M.E. Trenkel, Slow- and Controlled-Release and Stabilized Fertilizers: An Option for Enhancing Nutrient Use Effciency in Agriculture, International Fertilizer Industry Association [IFA], Paris,France, 2010, pp. 1—133.

[108] Deleted in review.

[109] Deleted in review.

[110] W. Vendrame, K.K. Moore, T.K. Broschat, Interaction of light intensity and controlled-release fertilization rate on growth and flowering of two new Guinea impatiens cultivars, HortScience 14 (2004) 491—495.

[111] Deleted in review.

[112] C.Y. Wang, Syntheses, crystal structures, and urease inhibitory properties of copper [II] and zinc[II] complexes with 2-bromo-4-chloro-6-[[2-dimethylaminoethylimino] methyl]phenol, J. Coord. Chem. 62 (17) (2009) 2860—2868, https://doi.org/10.1080/00958970902946702.

[113] Q. Wang, Y.L. Wang, J.L. Guo, C.X. Guo, Z.P. Yang, Release of sulfur coated urea in mild saline-alkali soil and study of its fertilizer efficiency, Acta Agric. Boreali-Sinica 31 (2016a) 182—187.

[114] Deleted in review.

[115] X.W. Wang, J.L. Kuai, J.H. Yu, X.J. Liu, Effects of controlled/slow-released nitrogen fertilizers on physiological characteristics and quality of melon under substrate cultivation, J. Plant Nutr. Fertil. 22 (2016b) 847—854.

[116] Deleted in review.

[117] D.P. White, Survival, growth, and nutrient uptake by Spruce and Pine seedlings as affected by slow-release fertilizer materials, in: C.T. Youngberg (Ed.), Forest-soil Relationships in North America, Oregon State University Press, Corvallis, 1965, pp. 47–63.

[118] C.F. Yamamoto, E.I. Pereira, L.H.C. Mattoso, T. Mattoso, C. Ribeiro, Slow release fertilizers based on urea/urea–Formaldehyde polymer nanocomposites, Chem. Eng. J. 287 (2016) 390–397, https://doi.org/10.1016/j.cej.2015.11.023.

[119] H.Z. Yang, G.L. Zheng, H.L. Liu, Q.H. Lin, W. Luo, Types of slow/controlled release fertilizer and its quality evaluation method, Chin. J. Tropical Agric. 36 (2016) 21–27.

[120] Y. Yang, X.Y. Ni, Z.J. Zhou, L.X. Yu, B.M. Liu, Y. Yang, Y.J. Wu, Performance of matrix-based slow- release urea in reducing nitrogen loss and improving maize yields and profits, Field Crop. Res. 212 (2017) 73–81, https://doi.org/10.1016/j.fcr.2017.07.005.

[121] L.L. Zhang, Z.J. Wu, L.J. Chen, Effect of coating and hydroquinone incorporation on urea-N release and its hydrolysis, Ecol. Environ. Sci. 18 (3) (2009) 1112–1117.

[122] B. Zhang, J. Gao, Y. Chen, A.E. Li, Creamer, H. Chen, Slow-release fertilizers encapsulated by graphene oxide films, Chem. Eng. J. 255 (1) (2014) 107–113.

[123] Q.W. Zhang, F. Saito, A review on mechanochemical synthesis of functional materials, Adv. Powder Technol. 23 (2012) 523–531.

[124] T. Zhou, Y. Wang, S. Huang, Y.C. Zhao, Synthesis composite hydrogels from inorganic-organic hybrids based on leftover rice for environment-friendly controlled-release urea fertilizers, Sci. Total Environ. 615 (2018) 422–430, https://doi.org/10.1016/j.scitotenv.2017.09.084.

CHAPTER 5

Methods for controlled release of fertilizers

Vinaya Chandran, Hitha shaji, Linu Mathew
School of Biosciences, Mahatma Gandhi University, Kottayam, Kerala, India

1. Introduction

Controlled-release fertilizers (CRFs) are fertilizer granules encapsulated within carrier molecules which control the release of fertilizer to agricultural field [1]. These encapsulated carrier molecules are usually known as excipients that supply nutrients to the crop based on their needs. CRF is one of the most efficient and advanced technologies being used throughout the world. It enriches soil nutrient level in a controlled manner and eliminates most of the negative impacts of traditional fertilizers. CRF is a good and alternative way to increase fertilizer usage efficiency (NUE), especially nitrogen in the soil [2].

CRFs can be partitioned into three classes depending on their excipients and supplement composition. The excipients play a key role in the gradual release of nutrients [3]. The other parameters controlling nutrient release include prill radius, soil temperature, moisture content, and soil microbial activity [4–6]. Fertilizer manufacturers control the nutrient release by changing the coating thickness and coating composition of CRF. In addition, nutrient composition may differ among different manufacturers. Thus, a series of CRFs are now available with a range of nutrient release characteristics.

- Uncoated nitrogen–based fertilizers
- Coated nitrogen–based fertilizers
- Polymer–coated controlled–release fertilizers

2. Uncoated nitrogen-based fertilizers

This is the oldest form of CRF consisting of chemically bound urea. These products partially reduce nutrient leaching. The release rate of uncoated nitrogen-based fertilizers is determined by particle size, availability of water, and microbial decomposition [7]. Urea formaldehyde (UF) is the first

Controlled Release Fertilizers for Sustainable Agriculture
ISBN 978-0-12-819555-0
https://doi.org/10.1016/B978-0-12-819555-0.00005-4

uncoated nitrogen–based fertilizer used for the slow release of N [8,9]. This product is made by reacting excess urea under controlled conditions such as temperature, pH, urea-formaldehyde ratio, and reaction time. The UF is a blend of unreacted urea, dimers, and oligomers such as monomethylol urea, dimethylol urea (DMU), and methylene urea. Finally, the addition of acid to this product results in the production of less water-soluble UF fertilizer [10,11]. The working mechanism of UF mainly depends on microbial action. Other factors such as soil pH, biological activity, clay content, moisture content, drying, wetting, and temperature may be involved in UF decomposition [8,10–12].

2.1 Isobutylidene diurea (IBDU)

IBDU is an example for uncoated nitrogen-based fertilizer. IBDU is a mixture of liquid isobutyraldehyde and solid urea. Here, chemical decomposition or hydrolysis plays a key role in N release. Thus, particle size of the fertilizer, soil moisture content, temperature, and pH are strongly involved in N release rate of uncoated nitrogen-based fertilizers [11].

2.2 Crotonylidene diurea (CDU)

CDU is made by reacting acetaldehyde with urea under the catalysis of an acid. The nitrogen is released to the soil or medium mainly through a combination of microbial degradation and hydrolysis [11].

3. Coated nitrogen-based fertilizers

Sulfur-coated urea (SCU) was one of the first CRFs. It was manufactured by the Tennessee Valley Authority [13,14]. SCU was prepared by coating preheated urea granules with elemental sulfur called molten sulfur. Sulfur is a low-cost plant macronutrient and is suitable for fertilizer coating because of its ability to melt at about 156°C. Jarrell et al. [15] describe that a typical SCU granule was made up of three types of coatings. The basic coating was enabled by spraying with molten sulfur over urea. Subsequently, a wax sealant is also sprayed to seal crack in the coating. The wax sealant acts as a perfect protective layer and minimizes microbial degradation [16–18]. Finally, a third layer (usually attapulgite) is added, which functions as a conditioner. The release of N from SCU is controlled by the thickness and coating quality [7]. SCU coated with perfect and thick coating is referred as "locked-off" [19] and damaged coating is denoted as "failure release" [20].

A batch of SCU granules may contain more than 33% damaged granules and rest "perfectly coated" granules [21,22]. Hence, 33% or more of the SCU content may be immediately discharged or burst after contact with water and remaining may be released long after (the "lock off" effect) if it is needed by the plant [7,23].

4. Polymer-coated controlled-release fertilizer

Polymer-coated CRFs have broad use in agriculture since they can be intended to release nutrients in a progressively controlled way. Different types of polymers have been used in fertilizer coating. Such polymers could be thermoplastic, thermosetting, or biodegradable ones. These are generally long-lasting and show stable, predictable release rates under average temperature and moisture conditions. Mechanism of polymer-coated CRF is dependent on the composition or thickness of polymer coating. Several synthetic and natural polymers are used for coating fertilizers [24]. Synthetic polymer coatings are typically made of blends of water permeable and impermeable resins, and surfactants like ethylene vinyl acetate, and talc occur as layered plates [25]. Here, nutrient release is not much affected by soil properties (pH, salinity, texture, microbial activity, redox-potential, and ionic strength of the soil solution). But the temperature and moisture permeability of the polymer coating are significantly important in nutrient release [26,27]. PCRF has more advantages than the previous category CRFs. It is available as mixtures of element formulations including N, P, K, and other formulations including micronutrients. Individual particles or granules of PCRF are called as prills. It consists of soluble nutrients within a permeable/semipermeable coating. There are two main groups of polymer-coated fertilizers depending on the coating materials. They are
- sulfur + polymers, together with wax polymeric materials
- polyolefin/polymeric materials
  Normally used coatings are
- polymers (e.g., polyolefin, polyethylene, polyvinylidene chlorid (PVDC)-based copolymers, gel-forming polymers, polyesters ethylene-vinyl-acetate, urea formaldehyde, resin, polyurethanelike resins, alkyd-type resins, etc.),
- fatty acid salts (e.g., calcium stearate),
- latex, rubber, guar gum, petroleum-derived anticaking agents, wax,
- magnesium and calcium phosphates, magnesium oxide, magnesium potassium phosphate, and magnesium ammonium phosphate,

- phosphate rock, phosphor gypsum, attapulgite clay,
- peat (encapsulating within peat pellets: organomineral fertilizers),
- neem cake/'nimin"-extract (extract from neem cake).

4.1 Polymer-coated SCU

Polymer-coated sulfur-coated fertilizers are modified products of SCU. Here SCU is coated with an additional thin coating of an organic polymer (thermoplastic or resin). This modified form is also called hybrid SCU. The release mechanism of PSCU is same as common sulfur-coated CRFs [19]. The added polymer layer provides an extra resistance to the coated granules, which in turn offers a better release performance than the SCU.

4.2 Polyolefin-coated urea (POCU)

POCU is a temperature-dependent controlled-release granular fertilizer. The coating or excipients of POCU is made by a blend of polyolefin, ethylene vinyl acetate, and talc as major components. The dissolution rate of this fertilizer granules doubles for every 10°C increase in temperature [24]. Polyolefin-coated urea fertilizer (POCU) was first manufactured in Japan [24] and a number of POCU with varying release rates are commercially available in Japan. By changing resin-talc ratio, different products with different dissolution rates could be obtained. Nutrient release primarily depends on temperature, which in turn modulates membrane permeability of resins coating. The mechanism of POCU involves water movement into the granule by osmotic potential, which dissolves urea inside the coating and diffuses out through the coating.

4.3 Resin-coated fertilizers

Coatings of Resin-coated fertilizers are made by in situ polymerization resulting in the formation of a cross-linked, hydrophobic polymer called resin. They are usually thermosetting, i.e., degrades upon heating. Some of the common thermoset resins include alkyd resin, epoxy resin, melamine resin, phenol resin, silicon resin, urethane resin, unsaturated polyester resin, and urea resin. Alkyd-type resins (e.g., osmocote) and polyurethane like coatings (e.g., polyon, plantacote, and multicote) are the leading groups of resins in practical use [8].

Thermoplastic coatings are made by dissolving the polymer coating material (e.g., polyethylene) in a chlorinated organic solvent and spraying the mixture on the granules in a fluidized bed reactor [28,29]. Nutrient release from thermoplastic coatings can be controlled and synchronized by manipulating the ratio of the coating components with different moisture permeability (e.g., ethylene–vinyl–acetate and polyethylene for Nutricote).

4.4 Biodegradable organic polymers

4.4.1 Pectin

Pectin is a complex heterogeneous, hydrophilic polysaccharide consisting mainly of esterified D-galacturonic acid residue and its methyl ester in α (1—4) chain found in the cell wall of higher plants. Certain fruits such as orange, apple, plum, quince, grapes, cherries are also known to contain the methoxyester of pectic acid. Due to its easy availability and low production cost, pectin is widely used in drug delivery. CRF based on pectin has an important application in removing Pb^{2+} and Cu^{2+} from wastewater and water and also in the release of urea, potassium, and phosphate. From this finding it is revealed that the pectin-based hydrogel is ideal for horticultural plants for effective conserving of water [30]. Depending on the plant source and preparation, pectin possesses varying degrees of methyl ester substituent, which is crucial for determining its solubility and requirements for gelation, for example, low methoxy pectin (LM) is more hydrophilic and soluble in pH 7.4 buffer than high methoxypectins (HM) which is poorly soluble and forms gel at pH around 3. Chemical modification such as saponification catalyzed by bases, mineral acids, salts of weak acids, enzymes, concentrated ammonium systems, and primary aliphatic amines also reduces the high solubility of pectin without changing their biodegradability [31].

4.4.2 Tamarind seed polysaccharide (TSP)

It is commonly called as galactoxyloglucan; a monomer, of mainly three sugars such as glucose, xylose, and galactose formulated in molar ratio of 3:2:1 isolated from kernel seed of *Tamarindus indica*. TSP is a natural biopolymer. It is nontoxic, cheap, and biocompatible. TSP-based agro material used in CRF practices gives effective chemical delivery systems [1]. It was observed that the increased concentration of TSP shows a maximum hardness and minimum friability, thus decreases the release rate; for example, TSP tablets of 20% showed maximum chemical delivery compared to tablets of 40% in a 24 h time period. TSP tablets can be used for the slow release of both water-soluble and water-insoluble chemicals [32].

4.4.3 Mimosa pudica seed mucilage

For sustainable delivery of formulations *Mimosa pudica* mucilage acts as a matrix forming agent. The sustained release depends on the concentration of mucilage and the chemical ratio. The diffusion of chemical through the gel layer is retarded, as and when the concentration of the mucilage increases. This is because it increases the gel strength and thereby reduces the diffusion coefficient of the nutrient. Lowering the concentration of the mucilage facilitates the release of chemical by diffusion and erosion mechanism, while higher concentration releases the chemical by means of diffusion through the matrix that is swollen [33].

4.4.4 Guar gum

It is a naturally occurring nonionic hydrophilic polysaccharide obtained from the seeds of *Cyamopsistetra gonolobus*. This binder or disintegrant is used in a solid dosage form, which possesses a release hindering property and is susceptible to microbial degradation. In presence of cold water, it swells and forms a sol or viscous colloidal dispersion. This gelling property helps to retard the release of chemical from the dosage tablet form [7]. The increased concentration of guar gum decreases the drug release whereas the increased gum concentration raises the swelling index value, thereby, resulting in slow erosion of gelled layer which in turn favors slow release.

4.4.5 Terminalia catappa gum (TC)

It is a natural release retarding polymer gum. The exudate is obtained from *Terminalia catappa* Linn. Kumar et al. [34] demonstrated the excellent swelling properties of TC gum in water and its ability to control the release of dextromethorphan hydrobromide from matrix tablet. Thus, the tablet formulations containing TC gum as an excipient may ensure controlled drug delivery systems of sparingly water-soluble, low molecular weight drug substances.

4.4.6 Gellan gum

It is a high molecular weight, hydrophilic anionic deacetylated exocellular polysaccharide gum isolated from a pure culture of *Pseudomonas elodea* as a fermentation product. Gellan Gum consists of a tetrasaccharide repeating unit of one α-L-rhamnose, one β-D-glucuronic acid, and two β-D-glucose residues. It is a natural, hydrophilic polysaccharide obtained from the inner bark of the tree *Grewia mollis,* which is known to hydrate on contact with water and swells to form a highly viscous dispersion making them very suitable for CRFs [35].

4.4.7 Mucuna gum

Mucuna gum is a biodegradable, amorphous polymer having emulsifying and suspending properties. The polysaccharide gum composed of mainly D-galactose along with D-glucose and D-mannose is isolated from the cotyledon of edible bean, *Mucuna flagillepes*. Udeala and Uwaga [36] showed that formulation without having a cross-linking showed the highest drug release. This signifies the use of mucuna gum as CRFs exhibiting similar features.

4.4.8 Gum copal (GC) and gum dammar-GD

It is a naturally occurring hydrophobic resin isolated from the plant *Bursera bipinnata*. Gum dammar is also anticipated to exhibit similar release kinetics as in GC. GD is a naturally occurring hydrophobic gum obtained from plant *Shorea wiesneri*. Morkhade et al. [37] evaluated natural gum copal (GC) and gum dammar (GD) as a novel sustained release matrix forming materials in tablet formulation. Along with the physicochemical properties, gum copal GC and gum dammar GD were also characterized for their polydispersity index, molecular weight, and glass transition temperature. The matrix tablets are prepared by wet granulation technique by using a granulating agent isopropyl alcohol. Matrix tablets made with 30% w/w of gum copal and gum dammar showed a controlled drug delivery beyond 10 h. The model drugs used are hydroxy propyl methyl cellulose and diclofenac sodium which are hydrophilic. The influence of concentration of hydroxyl propyl methylcellulose on drug release pattern of hydrophobic material was determined by forming an optimum ratio of drug:polymer 1:1. It was found that the hydrophobic:hydrophilic polymer ratio of 75:25 shows a similar release pattern. Hence at this ratio for hydrophobic matrices the initial burst-release was lowered to a great extent [38].

4.4.9 Karaya gum or xanthan gum

This is a hydrophilic naturally occurring gum obtained from *Sterculia urens* and composed of rhamnose, galactose, and glucuronic acid. Xanthan gum or karaya gum was directly used as matrices for producing release-controlling agent. This swells in water and is thus used as release rate controlling polymer in different formulations of two model drugs, diclofenac sodium and caffeine. These drugs possess different solubilities in aqueous medium [39]. Xanthan gum or karaya gum displayed a high degree of swelling due to increased water uptake and polymer relaxation aiding in small degree of erosion. In conclusion, chemical release from xanthan and karaya gum matrices depends on solubility, agitation speed, and proportion of the chemical [40].

4.4.10 Gum acacia

This soft–gelatin mainly used as encapsulating agent; it is isolated from stems of the *Acacia arabica* tree. Different gelatin composition has been found. Earlier compositions comprised of a nonbrittle shell forming material. Later they consisted of wet composition comprising 30%–60% by weight of a film-forming material, 5%–35% by weight of a water-dispersible or water-soluble plasticizer, 25%–65% by weight purified water. The film-forming material comprises of gum acacia and gelatin [41]. Other gums in the list include tragacanth gum used for sustained release, khaya gum from *Khaya grandifoliola* used as binding and coating agent, okra gum from *Hibiscus esculentus* used in the formulation of sustained-release [42] and *Hibiscus rosa sinesis* mucilage used to improve binding efficacy and act as release retarding agent.

4.4.11 Modified clays

Nanoclay the most common clay made by nanoparticles has been extensively used to produce CRFs. Compared to micrometer size materials nanolayers have large surface areas and greater reactivity. These interfaces and the surfaces provide an active substrate for biological, chemical, and physical reactions. These features provide a beneficial role for monolayer to be used as a reservoir of fertilizers or making sustainable carriers [1]. For the control of broadleaf weeds of different crops, herbicide atrazine was incorporated with ethyl cellulose for controlled-release formulations. Nanoclays are incorporated as matrix modifying agents. Clays and nano-clays used as matrix modifying agents showed little influence on the herbicide encapsulation and particle morphology. These particular factors show an advantage of CRFs in prolonged bioefficiency and minimize the harmful impact on the environment by producing longer application and sustained release of fertilizers [43].

5. Commercially available polymer-coated CRF

5.1 Osmocote

Osmocote [44] is referred as the first resin-coated fertilizer manufactured in California in 1967. It was made up of an alkyd-type resin with a glycerol ester [11,45]. Resin is applied in a number of layers, and the coating thickness and composition regulates the nutrient release. The mechanism of nutrient release from osmocote is as follows: water enters to the excipients through the micropore and upturns the osmotic pressure inside the excipient core, which in turn creates stretching in coating. This stretching again increases the number of micropores and nutrients are released through them [46]. The nutrient release from osmocote products is mostly temperature dependent [22,47,48], while soil moisture content, wetting and drying, soil microbial activity, and soil pH have minute effect on the nutrient release [47,48]. Different forms of osmocote are commercially available with discharge periods from as short as 3—4 months to up to 14—16 months.

5.2 Nutricote

Nutricote and multicote [49] utilize thermoplastic resins as coating materials that are highly impermeable to water. Thus, the coating is blended with special release controlling agents such as ethylene-vinyl acetate and surfactants to attain required permeability and longevity. Here, coating thickness is same for all the formulations and the added release control agents determine nutrient release rate. A number of nutricote with different formulation and longevity (from 4 to 16 months) are now available.

5.3 Polyon

Polyon [50] uses a reactive layer coating (RLC) process that polymerizes two reactive monomers as applied over the fertilizer core in a continuous coating drum, resulting in the formation of an ultrathin polyurethane membrane coating. Apparently, the efficiency of the RLC process allows for somewhat lower production costs than many other PCF.

6. Preparations of CRF formulations

Encapsulation of conventional fertilizer is done by different physical methods. These methods include in situ polymerization, spray coating, simple mixing, spray drying, pan coating, rotary disk atomization, etc. Simple mixing is carried out by simply mixing the fertilizer granule with the coating material at its melting point or with polymer solution in a suitable solvent. Polymer solution with a suitable solvent is sprayed on fertilizer granules and dried to remove solvents by evaporation. This method is commonly used for fertilizer encapsulations and is called spray coating. The equipments for these methods are rotary drum, fluidized bed reactor pan, ribbon mixer, or paddle mixer [51].

7. Field application methods

For the ornamental and native nurseries, PCRF are applied in two ways: preincorporated in to the growing media at the time of planting, and as top dressings during the season. If it is applied as a top dress particularly, CRF is placed on a growing medium. Incorporation into growing media is another method of using PCRF in the smaller containers used in forest conservation and agricultural fields. During outplanting, PCRF can be placed under or near plants [52]. Researchers recommend placing PCRF at the bottom of the planting hole, which guarantees that released supplements will be easily reached to the plant [53]. Another method is applying the PCRF in a dibbled hole alongside the plant or broadcasting it around its base. After the fertilizer's application, water vapor infiltrates through the hydrophobic coating membrane into the granule. This mediates nutrient dissolution and arising osmotic pressure leads to either a partial tearing off of the coating or to its swelling, which allows ease of ion transport through the coating [54].

8. Biodegradability of CRFs

The term *biodegradability* means ability of material to decompose within a short time period. The factors that affect the biodegradation of coated polymers are morphology, chemical structure, molecular weight, physicochemical factors such as ionic strength and pH, mechanical stresses, etc. The nutrient release from the biodegradable polymers occurs as a result of hydrolysis of the polymeric chains into nontoxic type of erosion. The erosion is due to the hydrolysis or enzymatic action by microbes which start from the outside. A broad number of approaches are being used improve the biodegradable polymeric systems, for example, formation of cross-linked polymeric networks by hydrophilic copolymers (e.g., polyethylene glycol) with biodegradable polymers and the hydrogels [51].

9. Advantages of controlled release of fertilizers

Unlike soluble conventional fertilizer CRFs release nutrients for an extended period. These are released gradually throughout plant's life or a season [55]. Thus, a single application with CRF can reduce manual labor and thereby save fertilizer cost [56,57]. CRF ensures nutrient availability by the controlled release of fertilizers and reduces stress and toxicity. The supply of CRFs enhances seed germination and quality crop growth with reduced disease infestation and lodging (undesired plant growth) [8,46,58]. Compared to traditional fertilizers, CRF overcomes nutrient leaching and ground water pollution [59]. Nutrient leaching to the environment depends upon their concentration in soil. CRFs maintain the nutrient level by gradual releasing of fertilizer. Thus, there is nothing surplus to lose to the environment. With a solitary application of CRF, the crop yield is equal or greater than that acquired with numerous applications of soluble N fertilizer [60].

10. Drawback of controlled-release fertilizers

Releasing properties of CRFs are governed by permeability or solubility of coating materials, microbial decomposition, and other environmental factors such as temperature, soil pH, and moisture content. Nutrient requirement, time periods of nutrient uptake, duration of plant growth, and preferred chemical forms are specific to each plant variety. Therefore, CRF selection for a particular crop variety may possibly depend on the above factors, for example, selection of a CRF with too low release rate may result in slow nutrient discharge causing nutrient deficiencies, retarded plant growth, and reduced yield. Similarly, excess CRF with high nutrient release rate may result in plant toxicity and injury, and the loss of CRF benefits [61]. Some polymer-coated products present a dilemma of persistence in the soil of the synthetic coating material [62]. It is one of the major drawbacks of polymer-coated CRFs. A good alternative to this problem is the production of CRFs with biodegradable materials as coatings [63].

11. Conclusion

CRFs are greatly beneficial to agriculture, horticulture, and silviculture. This technology has much economical, ecological, and environmental advantage than the conventional fertilizers, but their use is still limited. It is estimated that the CRFs represent a mere 1% of the total quantity of fertilizers. The enormous potential of CRFs can be fully realized only by solving issues related to their development, production, and application. The high costs of CRFs should also be addressed. Biodegradable coatings and development of new methods for controlled release are to be considered in this regard. Environmental impact of the newly developed CRFs should also be assessed.

References

[1] S.I. Sempeho, H.T. Kim, E. Mubofu, A. Hilonga, Meticulous overview on the controlled release fertilizers, Adv. Chem. 2014 (2014).

[2] B. Zhao, S. Dong, J. Zhang, P. Liu, Effects of controlled-release fertiliser on nitrogen use efficiency in summer maize, PLoS One 8 (8) (2013) e70569.

[3] W. Li, L. Zhang, C. Liu, Z. Liang, Preparation and property of poly (acrylamide-co-acrylic acid) macromolecule slow-releasing fertilizer, Int. J. Electrochem. Sci. 7 (11) (2012) 11470−11476.

[4] C.W. Du, J.M. Zhou, A. Shaviv, Release characteristics of nutrients from polymer-coated compound-controlled release fertilizers, J. Polym. Environ. 14 (3) (2006) 223−230.

[5] L. Kaplan, P. Tlustoš, J. Száková, J. Najmanová, The influence of slow-release fertilizers on potted chrysanthemum growth and nutrient consumption, Plant Soil Environ. 59 (9) (2013) 385−391.

[6] L.C. Carson, M. Ozores-Hampton, Factors affecting nutrient availability, placement, rate, and application timing of controlled-release fertilizers for Florida vegetable production using seepage irrigation, HortTechnology 23 (5) (2013) 553−562.

[7] H.M. Goertz, Controlled release Technology, in: Kirk-Othmer Encyclopedia of Chemical Technology, vol. 7, Controlled Release Technology (Agricultural), 1993, pp. 251−274.

[8] M.E. Trenkel, Controlled-release and Stabilized Fertilizers in Agriculture, vol. 11, International fertilizer industry association, Paris, 1997.

[9] A. Shaviv, Preparation Methods and Release Mechanisms of Controlled Release Fertilisers, International Fertiliser Society, 1999.

[10] A. Alexander, H.U. Helm, Ureaform as a slow release fertilizer: a review, Z. für Pflanzenernährung Bodenkunde 153 (4) (1990) 249−255.

[11] H.M. Goertz, Commercial granular controlled release fertilizers for the specialty markets, in: R.M. Scheib (Ed.), Controlled Release Fertiliser Workshop Proceedings, 1991, pp. 51−68.

[12] T. Aarnio, P.J. Martikainen, Mineralization of C and N and nitrification in scots pine forest soil treated with nitrogen fertilizers containing different proportions of urea and its slow-releasing derivative, ureaformaldehyde, Soil Biol. Biochem. 27 (10) (1995) 1325−1331.

[13] G.M. Blouin, D.W. Rindt, U.S. Patent No. 3295950, U.S. Patent and Trademark Office, Washington, DC, 1967.

[14] S.P. Landels, Controlled-Release Fertilizers: Supply and Demand Trends in US Nonfarm Markets, Publisher: SRI International, Menlo Park, CA, USA, 1994.

[15] W.M. Jarrell, G.S. Pettygrove, L. Boersma, Characterization of the thickness and uniformity of the coatings of sulfur-coated urea 1, Soil Sci. Soc. Am. J. 43 (3) (1979) 602−605.

[16] S.E. Allen, C.M. Hunt, G.L. Terman, Nitrogen release from sulfur-coated urea, as affected by coating weight, placement and temperature[1], Agron. J. 63 (4) (1971) 529−533.

[17] R.M. Scheib, G.H. McClellan, Characteristics of sulphur texture on SCU, Journal (1976).

[18] W.M. Jarrell, L. Boersma, Release of urea by granules of sulfur-coated urea 1, Soil Sci. Soc. Am. J. 44 (2) (1980) 418−422.

[19] H.M. Goertz, Technology developments in coated fertilizers, in: Proceedings: Dahlia Greidinger Memorial International Workshop on Controlled/Slow Release Fertilizers, Technion-Israel Institute of Technology, Haifa, March 1993, pp. 7−12.

[20] S. Raban, E. Zeidel, A. Shaviv, Release mechanisms-controlled release fertilizers in practical use, in: J.J. Mortwedt, A. Shaviv (Eds.), Third Int. Dahlia Greidinger Sym. On Fertilisation and the Environment, 1997, pp. 287–295.

[21] J.D. Fry, O.L. Fuller, F.P. Maier, Unreleased nitrogen in sulfur-coated urea and reactive layers coated urea following application to truth, in: Controlled Release Fertilizer Workshop, TVA, 1991, pp. 40–43.

[22] S. Raban, Release Mechanisms of Membrane Coated Fertilizers. Research Thesis, Israel Institute of Technology, Haifa, Israel, 1994.

[23] A. Shaviv, S. Raban, E. Zaidel, Modeling controlled nutrient release from polymer coated fertilizers: diffusion release from single granules, Environ. sci. & technol. 37 (10) (2003) 2251–2256.

[24] A.T. Gandeza, S. Shoji, I. Yamada, Simulation of crop response to polyolefin-coated urea: I. Field dissolution, Soil Sci. Soc. Am. J. 55 (5) (1991) 1462–1467.

[25] S. Shoji, MEISTER Controlled Release Fertilizer: Properties and Utilization, Konno Printing Company Limited, 1999.

[26] S. Shoji, A.T. Gandeza, in: S. Shoji, A.T. Gandeza (Eds.), Controlled Release Fertilizers with Polyolefin Resin Coating, Konno Printing Co., Sendai, Japan, 1992, pp. 1–7.

[27] T. Fujita, S. Shoji, Kinds and properties of meister fertilizers, in: Di Dalam Meister Controlled Release Fertilizer—Properties and Utilization, Konno Printing Company Ltd, Sendai, Japan, 1999.

[28] T. Fujita, C. Takahashi, S. Yoshida, H. Shimizu, U.S. Patent No. 4369055, U.S. Patent and Trademark Office, Washington, DC, 1983.

[29] T. Fujita, S. Maeda, M. Shibata, C. Takahashi, Research & development of coated fertilizer, in: Proceedings of the Symposium on Fertilizer, Present and Future, Japanese Society of Soil Science & Plant Nutrition, Tokyo, 1990, pp. 78–100. September 25–26,.

[30] M.R. Guilherme, A.V. Reis, A.T. Paulino, T.A. Moia, L.H. Mattoso, E.B. Tambourgi, Pectin-based polymer hydrogel as a carrier for release of agricultural nutrients and removal of heavy metals from wastewater, J. Appl. Polym. Sci. 117 (6) (2010) 3146–3154.

[31] A. Singh, P.K. Sharma, R. Malviya, Release behavior of drugs from various natural gums and polymers, Polim. Med. 41 (4) (2011) 73–80.

[32] S. Kumar, S.K. Gupta, Natural polymers, gums and mucilages as excipients in drug delivery, Polim. Med. 42 (3–4) (2012) 191–197.

[33] K. Lubkowski, B. Grzmil, Controlled release fertilizers, Pol. J. Chem. Technol. 9 (4) (2007) 83–84.

[34] S.V. Kumar, D. Sasmal, S.C. Pal, Rheological characterization and drug release studies of gum exudates of Terminalia catappa Linn, AAPS Pharm. Sci. Tech. 9 (3) (2008) 885–890.

[35] T. Jamnongkan, S. Kaewpirom, Controlled-release fertilizer based on chitosan hydrogel: phosphorus release kinetics, Sci. J. Ubu. 1 (1) (2010) 43–50.

[36] O.K. Udeala, U.N. Uwaga, Some emulsifying and suspending properties of a polysaccharide gum derived from Mucuna flagillepes, Papilionaceae, J. Pharm. Pharmacol. 33 (1) (1981) 75–78.

[37] D.M. Morkhade, S.V. Fulzele, P.M. Satturwar, S.B. Joshi, Gum copal and gum damar: novel matrix forming materials for sustained drug delivery, Indian J. Pharm. Sci. 68 (1) (2006) 53.

[38] V.M. Fulbandhe, C.R. Jobanputra, K.J. Wadher, M.J. Umekar, G.S. Bhoyar, Evaluation of release retarding property of gum damar and gum copal in combination with hydroxypropyl methylcellulose, Indian J. Pharm. Sci. 74 (3) (2012) 189.

[39] O.A. Odeku, J.T. Fell, In-vitro evaluation of khaya and albizia gums as compression coatings for drug targeting to the colon, J. Pharm. Pharmacol. 57 (2) (2005) 163−168.

[40] D.L. Munday, P.J. Cox, Compressed xanthan and karaya gum matrices: hydration, erosion and drug release mechanisms, Int. J. Pharm. 203 (1−2) (2000) 179−192.

[41] A. Gennadios, Gum acacia substituted soft gelatin capsules, Banner Pharmacaps, United States Patent 6193999, 2001.

[42] V.D. Kalu, M.A. Odeniyi, K.T. Jaiyeoba, Matrix properties of a new plant gum in controlled drug delivery, Arch Pharm. Res. (Seoul) 30 (7) (2007) 884−889.

[43] M. Cea, P. Cartes, G. Palma, M.L. Mora, Atrazine efficiency in an andisol as affected by clays and nanoclays in ethylcellulose controlled release formulations, Rev. Ciencia del Suelo y Nutrición Veg. 10 (1) (2010) 62−77.

[44] Scotts Horticulture, Controlled Release Fertilizers, 2008. http://the-scotts-exchange. com/products/fertilizers/osmocote.cfm (Accessed 3 December 2008).

[45] J.M. Lambie, U.S. Patent No. 4657576, U.S. Patent and Trademark Office, Washington, DC, 1987.

[46] R.D. Hauck, Slow-release and Bioinhibitor-Amended Nitrogen Fertilizers, Fertilizer technology and use, fertilizertechn), 1985, pp. 293−322.

[47] C.B. Christianson, Factors affecting N release of urea from reactive layer coated urea, Fert. Res. 16 (3) (1988) 273−284.

[48] W.P. Moore, Reacted layer technology for controlled release fertilizers, in: Proceedings: Dahlia Greidinger Memorial International Workshop on Controlled/Slow Release Fertilizers, Technion-Israel Institute of Technology, Haifa, March 1993, pp. 7−12.

[49] Scotts Horticulture, Nutricote controlled release fertilizer, 2008. http://www.sungro. com/products_displayProBrand.php?brand_id=6 (Accessed 3 December 2008).

[50] PURSELL Inc, POLYON Polymer Coatings and the R/LCTM Process, Publisher: PURSELL Industries, Inc., Sylacauga, Alabama, USA, 1995.

[51] V. Madhavi, G. Madhavi, A. Reddy, A scrupulous overview on controlled release fertilizers, Agric. Allied Sci. 5 (2016) 26−33.

[52] D.F. Jacobs, R. Rose, D.L. Haase, Incorporating controlled-release fertilizer technology into outplanting, in: L.E. Riley, R.K. Dumroese, T.D. Landis (Eds.), National Proceedings: Forest and Conservation Nursery Associations-2002, US Department of Agriculture Forest Service Rocky Mountain Research Station Proc. RMRS-P-28, May 2003, pp. 37−42.

[53] J.F. Gleason, M. Duryea, R. Rose, M. Atkinson, Nursery and field fertilization of 2+ 0 ponderosa pine seedlings: the effect on morphology, physiology, and field performance, Can. J. For. Res. 20 (11) (1990) 1766−1772.

[54] M. Kochba, S. Gambash, Y. Avnimelech, Studies on slow release fertilizers: 1. Effects of temperature, soil moisture, and water vapor pressure, Soil Sci. 149 (6) (1990) 339−343.

[55] R.P. Wiedenfeld, Rate, Timing, and Slow-Release Nitrogen Fertilizers on Cabbage and Onions (No. RESEARCH), 1986.

[56] S. Shoji, A.t. Gandeza, Utilization of polyoelfin fertilizers to crops,In.Controlled release fertilizers with polyoelfin coatings, in: S. Shoji, A.T. Gandaza (Eds.), Development,Properties and Utilosation, Konno Printing Ltd, Sendai, Japan, 1992, pp. 37−42.

[57] A. Shaviv, Advances in Controlled-Release Fertilizers, 2001, pp. 1−49.

[58] E.A. Seward, Slow-Release Nitrogen Fertilizers, Nitrogen in crop production, (nitrogenincropp), 1984, pp. 195−206.

[59] F.L. Wang, A.K. Alva, Leaching of nitrogen from slow-release urea sources in sandy soils, Soil Sci. Soc. Am. J. 60 (5) (1996) 1454−1458.

[60] D. Drost, R. Koenig, T. Tindall, Nitrogen use efficiency and onion yield increased with a polymer-coated nitrogen source, Hortscience 37 (2) (2002) 338–342.

[61] A. Shaviv, Plant response and environmental aspects as affected by rate and pattern of nitrogen release from controlled release N fertilizers, in: Progress in Nitrogen Cycling Studies, Springer, Dordrecht, 1996, pp. 285–291.

[62] M. Kolybaba, L.G. Tabil, S. Panigrahi, W.J. Crerar, T. Powell, B. Wang, Biodegradable polymers: past, present, and future, in: ASABE/CSBE North Central Intersectional Meeting, American Society of Agricultural and Biological Engineers, 2006, p. 1.

[63] S.M. Al-Zahrani, Controlled-release of fertilizers: modelling and simulation, Int. J. Eng. Sci. 37 (10) (1999) 1299–1307.

CHAPTER 6

Manufacturing of slow and controlled release fertilizer

Chandra Wahyu Purnomo[1], Hens Saputra[2]

[1]Chemical Engineering Department, Universitas Gadjah Mada, Sleman, Yogyakarta, Indonesia; [2]Agency for the Assessment and Application of Technology, Jakarta, Indonesia

1. Introduction

Fertilizer is the common term used for materials that are intentionally added into the soil or on the plant's surface to fulfill its nutrient demand during the growing period. It is derived from a wide variety of natural and synthesized materials and commonly available in solid or liquid form. Commercial fertilizers commonly consist of three major nutrients, which are nitrogen (N), phosphorus (P_2O_5), and potassium (K_2O) denoted as N−P−K.

There are several classifications of fertilizer that related to its nutrient content, releasing pattern and origin. In terms of nutrient content, some contain only N source, while the complex ones contain N−P−K. A fertilizer is said to be complete or mixed when it contains nitrogen, phosphorus, and potassium as its primary nutrients in a certain ratio for instance: 10−10−10, 15−15−15, or 20−10−10. However, an incomplete fertilizer excludes one or more of the major components such as 46−0−0 (urea), 18−46−0 (diammonium phosphate), and 0−0−60 (potash).

In terms of nutrient origin, there are organic and synthetic fertilizers. The word organic means that the nutrients contained in the product are derived solely from the remains of a once-living organism. Meanwhile, synthetic fertilizer is manufactured from natural gas or minerals with specific processing steps. However, most organic fertilizer consists of lower nutrient content than the commercial ones.

The nutrient release rate of some fertilizers has been modified to provide targeted plants with a steady supply of nutrients throughout their most active periods of growth. Slow release fertilizers are designed to release nutrients at rates matching the specific plant requirements over an extended period. Its main purpose is to fulfill the increasing agriculture production demand without sacrificing resources and the environment.

Controlled Release Fertilizers for Sustainable Agriculture
ISBN 978-0-12-819555-0
https://doi.org/10.1016/B978-0-12-819555-0.00006-6

Fertilizers are considered as the most effective way to increase the crop yield in modern agriculture, especially for macronutrients, such as nitrogen (N), phosphorus (P), and potassium (K). However, there is a limit in plant uptake efficiencies, leading to serious environmental problems such as greenhouse gas production, water pollution, and eutrophication [17]). In addition, the low efficiency of fertilizer was caused by losses of nutrients to air, ground and surface water, as well as fixation of nutrients by soil.

Crop recovery of soluble fertilizer N is less than 50% and, however, has been potentially developed to minimize N losses through volatilization, denitrification, and leaching. It is expected that the controlled release fertilizer (CRF) can manage the release of nutrients and increase nutrient uptake efficiency (NUE). In Indonesia, the demand for chemical fertilizer for the rice field is ca. 200–400 kg/ha, with a reduced rate of nitrogen application due to the use of urease and nitrification inhibitors, thereby improving its efficiency.

Climate change has a substantial impact on the application of fertilizer and its efficiency, due to its effect on agricultural production activities and adaptation. Furthermore, rising temperatures and changes in precipitation lead to the volatilization or loss of soil minerals. In addition, temperature changes affect the biophysical and chemical processes of soil, thereby affecting its efficiency [3]).

The slow or controlled release of fertilizers is the best solution to some environmental problems caused by conventional chemical types. The NUE is improved according to the nutrients need in the growth and productive periods of plants. There are many advantages associated with this fertilizer type such as the ability to decrease loss rate, supplying nutrients sustainably, reduce the frequency of application, and minimize potential negative effects associated with overdosage. In general, slow release fertilizer (SRF) and CRF are interchangeable, with CRF is more sophisticated than SRF in controlling the nutrient release by Ref. [8]:

- less than 15% nutrient released in 1 day,
- 15%–75% nutrient released in 28 days, and
- at least 75% nutrient released in stated release time.

According to Ref. [14]; the slow release of fertilizers is generally classified into four types: (1) inorganic materials of low solubility, such as metal ammonium phosphates; (2) chemically or biologically degradable low solubility materials, such as urea-formaldehyde, (3) relatively soluble materials that gradually decompose in the soil, (4) water-soluble fertilizers controlled by a physical barrier, such as coated and matrix materials used to control the release of nutrient by the adsorption—desorption mechanism. In this chapter, the last type is further discussed due to its current popularity.

2. Coated fertilizer manufacture

The major categories of the controlled release fertilizers are coated fertilizers that are prepared by coating the fertilizer granules with various materials. Many materials have been reported to be used as coatings, such as sulfur, polysulfone, polyvinyl chloride, and polystyrene. The drawback of these fertilizers besides expensive is the soil accumulation of remaining coating materials to form a new type of pollution [10].

Various types of natural and biodegradable coating material such as starch and chitosan to nondegradable polymers have been utilized. Most of the commercial CRFs in the global market make use of this method in their production line with the total capacities of coated fertilizer estimated more than 3,000,000 tons in 2017 [1]. The mechanism of nutrient release from the core to the environment is diffusion through the thin coating materials after being penetrated by water. Therefore, it is essential to cover the entire surface evenly with the coating to avoid a sudden dissolution. In addition, the coating has to be able to withstand the swelling of the core due to water penetration and fertilizer dissolution. The criteria for selecting a coating material are low cost, nontoxic, and easy to be applied to conventional fertilizers.

There are two common types of equipment used to prepare coated fertilizers namely drum and spouted machines. The coating drum is commonly used for sulfur-coated fertilizer, while the spouting method is coated with a polymer. The manufacturing process is discussed more compared to the various types of coating materials.

2.1 Sulfur-coated fertilizer

One of the early attempts in adding a release barrier to commercial fertilizer is by sulfur coating. Sulfur is chosen due to its degradability and because it is considered as one of the plant nutrients. In the 1960s, some patents explained the production methods of sulfur-coated fertilizers with a patent from Tennessee Valley Authority (TVA) used to describe the manufacturing process of sulfur-coated urea (SCU) as shown in Fig. 6.1 (Patent No. 3,903,333, 1975) [15].

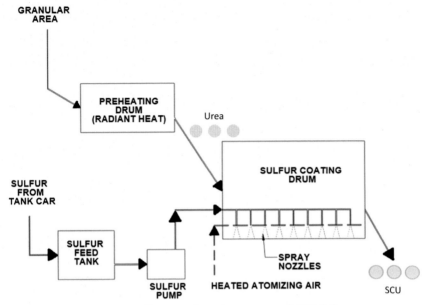

Figure 6.1 The manufacturing process of SCU [15].

In general, molten sulfur spray in a coating drum was used in the production of SCU. Before the coating process, the urea granule is preheated in another preheater drum for uniformity at a temperature range of 50–90°C.

The molten sulfur spray system needs to be properly built and controlled during operation, as the thin uniform coat on the urea surface is the key to SCU production. Therefore, the ununiformed coating is unwanted as the urea can burst at the uncoated spot during fertilizer application. The molten urea is sprayed at the temperature of 130–150°C by using steam as the heat source.

The other important operating condition that has to be closely controlled is the drum temperature. An electric heater is used to maintain the coating drum at the precise temperature of 100°C to allow sulfur solidification and optimum tensile strength. It is also important to conduct a better cooling mechanism after coating for longer nutrient release. A fast cooling mechanism is provided by fluidized cooling system.

Recently, additional precoat and postcoat techniques in SCU production were implemented to minimize the unevenly coated and surface crack granules that lead to a sudden burst of nutrient application. Initially, urea granules were precoated with an impervious sealant from a petroleum by-product such as petrolatum, motor oil, and soft wax. After precoating, the same procedure of sulfur coating is carried out. Finally, the SCU was subjected to a third coating procedure by using plasticizers such as polyethylene or polyvinyl acetate to aid the uniform spreading and fusion of the sulfur layer and avoid crack formation.

2.2 Polymer coated fertilizer

The extensive list of polymeric materials for fertilizer coating is found in several review papers [2]. However, this section is focused on the manufacturing step using a patent from Chisso Corporation Japan in 1991, which is a method of preparing coated fertilizer with more robust machinery as shown in Fig. 6.2 (Patent No. 5,009,696, 1991) [6]. The coating material was claimed degradable by UV light (sunlight), which was a mixture of polyolefin and ethylene-vinyl acetate-carbon monoxide (C_2H_4-VAc-CO) copolymer. As mentioned before, when the coating material is difficult to be degraded in nature, it leads to the generation of a new type of pollution.

The manufacturing process utilizes spouting encapsulation apparatus with the granular fertilizer fed through the fertilizer-feeding port while a definite quantity of hot air is passed to form a spout. In addition, when the exhaust temperature reaches a certain degree, an encapsulating liquid is blown through a nozzle toward the spout in the form of a spray.

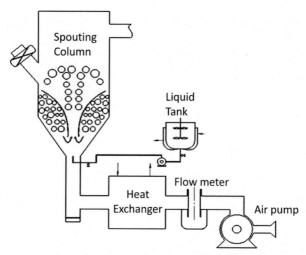

Figure 6.2 Spouting column configuration for producing polymer-coated fertilizers [6].

The encapsulating liquid contains solids content weighing 2.5%, with the liquid and powder sent into the tank through a pump, and the pipe that led to the nozzle is heated by steam. This prevents the temperature from decreasing to 100°C. The blower is stopped and the encapsulated fertilizer is withdrawn from a withdrawing port when the percentage encapsulation has reached a definite amount.

The development of this process changes the coating mixture with more degradable substances such as poly 3-hydroxy-3-alkylpropionic acid using a resin, or an inorganic or organic powder such as talc, $CaCO_3$, Al_2O_3 or white carbon as the filler (Patent No. 5,176,734, 1993) [7].

2.3 Hybrid coating

Some efforts have been conducted to combine the coating of sulfur with a polymer. This combination can improve the impact resistance of sulfur-coated fertilizer with the ability to reduce the thickness of polymer coating that leads to lower production costs. The early effort evolved from the SCU production after sulfur coating, due to the spray of some monomer mixtures of thermoset polymer. The liquid monomers are diisocyanates, such as 4,4-diphenylmethane diisocyanate (MDI), polyol mixture of diethylene glycol (DEG), and triethanolamine (TEA) (Patent No. 5,599,374, 1997) [5]. This improved procedure requires one additional spray drum after the sulfur coating drum with a special arrangement of spray nozzles. The diisocyanate has to be introduced separately from the polyol mixture, with the nozzle arrangement shown in Fig. 6.3.

The figure shows that a more complex combination of sulfur and polymer coating has been prescribed. The controlled release fertilizer composition comprises a water-soluble fertilizer core, coated with

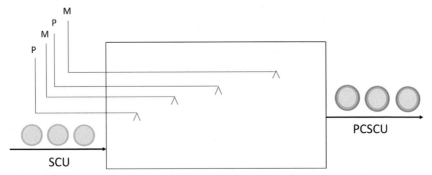

Figure 6.3 After the coating drum of SCU with polymer using a sequential nozzle [5].

polymeric, intermediate, and sulfur layers patented by Agrium Inc. (Patent No. US 8,741,022 B2, 2014) [11]. For example, the urea granule is first coated with sulfur followed by thermoset polymer (DEG-TEA polyol and diisocyanate, MDI) and wax. Several other combinations are also possible, such as the urea granule with the first coating of thermoset polymer followed by a layer of sulfur and wax. Another formulation was the coating of thermoset polymer followed by a layer of wax and sulfur with each combination consisting of a unique dissolution rate profile.

2.4 Biobased coating

Generally, the biobased coating materials have captured much attention, due to their relatively low-cost, abundant, renewable, and environmentally friendly properties. These materials include chitosan, keratin, polyamino acids, cellulose, and starch that are used directly to release fertilizers slowly [9]. Among these materials, starch, extracted from corn, sweet potato, and wild acorns, is considered to be an abundant and renewable natural polysaccharide. This is attributed to its several attractive features, such as biodegradability, nontoxicity, low cost, easy accessibility, and modifiability. However, the unmodified starch for nonfood field applications is often limited because of its poor solubility, low mechanical properties, instability at high temperature, and pH.

Chemical modification of starch involves the reaction of hydroxyl groups in the anhydroglucose units. Modified starch with good degradation was applied to slowly release the fertilizers with carboxymethyl starch (CMS), which is an important derivative applied in large industrial sectors such as food, paper, textile, plastics and pharmaceuticals. Moreover, CMS is relatively renewable, biodegradable, and widely used in fertilization [17].

3. Matrix dispersed fertilizer

The second manufacturing method is much simpler due to the use of hydrophobic solid materials such as matrices. This is mixed and shaped with conventional fertilizer powder to form composite particles. However, when this matrix incorporated fertilizer is applied in the field with moisture contact, the nutrient released becomes complex with the simultaneous occurrence of particle swelling, diffusion through the porous matrix, as well as gradual particle disintegration and dissolution from the outer surface. These intertwined mechanisms usually create difficulties in controlling the release of nutrients; therefore, it is more convenient to mention this matrix dispersed fertilizer as SRF. In addition, its strength and dissolution rate have to meet a certain level to ensure the gradual release of the contained nutrients. This SRF has a relatively lower production cost than coated CRF, and it is easily tailored to meet the formulation for a specific crop. However, coated CRF has gained more attention and penetration into the international market than matrix-based CRF. The release of a more predictable nutrient release profile and better appearance are the main reasons for the higher acceptance of coated CRF.

This type of controlled release fertilizer is quite common especially in developing countries like Indonesia, due to its low production cost and the simple manufacturing steps. In addition, several commercial-scale plantations such as sugarcane and palm oil have used this type of fertilizer as shown in Fig. 6.4.

It can be seen from Fig. 6.4B that the particle of SRF is not evenly shaped. Although the size and shape are not regular, it is still applicable to the real scale. It seems that the quality control of SRF is not very strict or perhaps it is not necessary as long as the nutrient content is suitable to the plantation and the SRF is spreadable by the machine (Fig. 6.4A).

In general, the preparation method is conducted by dispersing fine fertilizer particles into a hydrophobic matrix. By using this method, it is expected that the matrices become a diffusion barrier of water in diluting the nutrient contained. The particle distribution inside the matrix is

(A)   **(B)**

Figure 6.4 Matrix-based NPK fertilizer applied widely in a sugarcane plantation in Indonesia.

inhomogeneous and the nutrient release is also affected by its disintegration, thereby making the profile difficult to be predicted. Therefore, it is convenient to call matrix dispersed fertilizer as SRF rather than CRF.

The choice of the matrix is quite diverse from several types of waxes, clays, natural zeolites, and charcoals. As its necessity is quite large as fertilizer particle "storage," the matrices composition tends to make up approximately 20%—50% of the total weight. This is very much different from the coating type of SRF, whereby the coat materials do not exceed 10% of the total weight.

The high content of matrices inside the product is a drawback with some benefits. The major drawback is, of course, the reduction of nutrients inside SRF up to a half of the origin fertilizer. Therefore, farmers need to apply matrix-based SRF in the field two times more than conventional fertilizer assuming the want to utilize the same dosage.

The advantage of matrix dispersed fertilizer beside ease of production is associated with its properties. For instance, the use of natural zeolite can increase the soil exchange capacity to the nutrients and also lowers its acidity. Furthermore, when charcoal is used, the char increases the carbon content of the soil while serving as the growth media for soil microbes [13].

3.1 Wax dispersed fertilizer

This fertilizer type uses a matrix in the form of melting solid during processing to incorporate fine particles followed by a fast solidification process by water quenching. The SRF also needs a solid with a low melting point such as wax because it is impossible to expose fertilizer to a very high temperature. Although various types of animal, vegetable, and mineral wax can be used, petroleum wax such as paraffin is preferred for SRF production (Patent No. 3,242,237, 1966) (Fig. 6.5) [4].

The process diagram of this SRF is shown in Fig. 6.5. The solid fertilizer component of the SRF is reduced to subdivide the fertilizer particles into smaller mesh sizes of 100. The fine powder is then mixed with melted wax in a tank and then by gravity flow, the mixture is transferred to coil heated pipes with a wire end. By some mechanical vibration, the droplet is formed and then it falls into the water chamber. The mixture is solidified instantly in the water and then sorted by the vibrating screen.

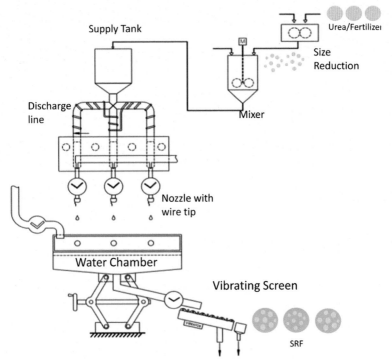

Figure 6.5 Wax dispersed fertilizer production method [4].

3.2 Natural material dispersed fertilizer

Natural materials such as zeolite, clay, charcoal, ashes, compost, and processed poultry manure are used as the SRF matrix. However, these materials do not require the melting process compared to conventional fertilizer. This type of matrices only requires room temperature mixing and shaping with binding properties such as clays or sludge. However, some materials need binder addition such as molasses or starch.

In general, the manufacturing process consists of pretreatment, formulation, granulation, and drying the product. Pretreatment of matrices is used for contamination removal and size reduction. Pretreatment especially size reduction is quite important to achieve a high quality of SRF product. The unwanted material can reduce the specification of zeolite especially the adsorption and cation exchange capacity (CEC). High adsorption capacity can increase CEC quality in the surrounding zeolite pore. Size reduction was conducted using a jaw crusher, with a hammer mill used to produce zeolite powder with a particle size of 60–100 mesh.

The selection of crusher's material construction is related to life time of equipment due to the high corrosive raw materials such as urea, potassium chloride, and diammonium phosphate.

After the pretreatment process, the matrices powder is mixed with fertilizer using a specific formulation. The mixture also considers the need for macro- and micronutrient for specific plants, while the generic SRF formula follows common multinutrient fertilizer or NPK with commercial ratio N–P–K are 20–10–10, 15–15–15, or 30–6–8.

After homogenous mixing step with a specific formulation, the forming method is conducted using a granulator, extruder, or tableting machine. The extruder system needs quite a lot of water to form an extrudable dough. The least water requirement uses tableting or pellet press machine, while a double roller pelletizer is one of the famous equipment in forming powder into pellets without water addition, thereby removing the dryer before packaging. The rotation of the rollers produces heat from the friction that dries the formed materials (Fig. 6.6).

The complete set of SRF manufacturing machinery is presented in Fig. 6.7. The formulation unit consists of several silos for storing nutrients and matrix with an automatic weighing system that produces formulation of multinutrient fertilizers such as NPK: 20–10–10 or 15–15–15. Furthermore, the measured quantity of each nutrient is transferred from the belt conveyor to the mixer. After the mixing process, the mixture is transferred to the crusher for size reduction, and then the fine powder to the pan or drum granulator. The pan granulation is utilized due to its low investment cost and the installation of several pans with a different fertilizer formulation. In this granulating process, a binder solution is added such as starch or molasses.

Some locally available materials can be used as SRF matrices such as poultry manure, and the production stage can be seen in Fig. 6.8 [12]. Poultry manure is dried, milled, and sieved to produce smell free and ready to be used manure powder for SRF preparation. The pellet is formulated using the granulator or screw extruder. Before pelleting, the manure powder is mixed with fertilizers and a binding material such as urea powder and starch, respectively. The mixture dough then extruded to produce noodle-like long shape mass, which is dried and cut at about 1 cm length of pellets. Meanwhile, SRF granule from poultry powder is prepared by mixing manure with urea powder and transferring the

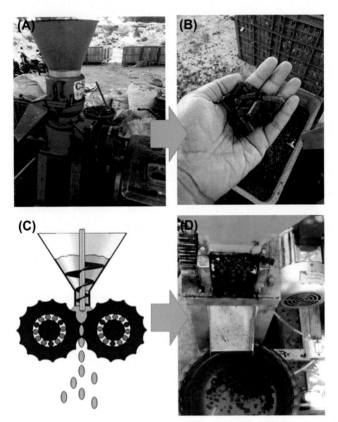

Figure 6.6 Fertilizer shaping machines, (A) roll and dices pelletizer, (B) the pellet product, (C) double roller press, (D) the roller press product.

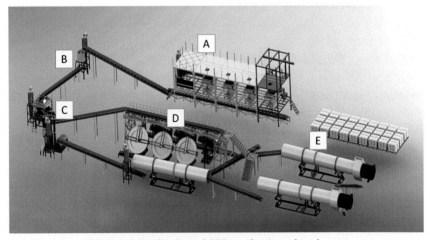

Figure 6.7 Zeolite-based SRF production plant layout.

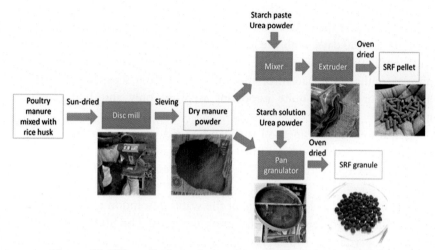

Figure 6.8 SRF production using poultry manure as the matrix [12].

mixture into the pan and rotated. During rotation, the binder solution is sprayed to produce a ball-shaped granule with the diameter range of 0.5—1.0 cm with the SRF granules dried and packed.

4. CRF testing procedure

CRF has no universal quality standard and characterization method; however, its common product testing includes dilution test, particle hardness, pH, and moisture absorption rate. The famous dissolution test was developed by TVA and was used to measure the amount of urea released by a 250 g coated sample immersed in 250 mL of water at 100°F for 7 days. The concentration of nutrients in demineralized water was analyzed periodically using 1 g of CRF immersed in 100 mL demineralized water with nitrogen. The total nitrogen obtained was analyzed using the Kjeldahl method.

The more robust study of nutrient release is the soil incubation test by immersing the sample of SRF in the soil for a certain time. Before the release study, the soil needs to be selected and pretreated. The selected soil type is collected from 0 to 20 cm depth and sieved passed through a 30-mesh sieve. It is further dried at room temperature to eliminate moisture for about a month. Subsequently, a soil sample weighing 100 g is transferred to a cylindrical pot with a diameter of 6.0 cm. SRF fertilizer is weighed and placed and mixed with soil in the pot, with some deionized water of approximately 38.69% added up to the common moisture content in the real field. The lid then placed on top of the pot and incubated for 1, 2, 3, 4, 6, 8, 10, and 60 days, with the ammonium and nitrate concentration, pH, and conductivity subsequently measured. In the soil test, the form of nitrogen either in the form of ammonia or nitrate is separately measured. This measurement is quite important as plants only absorb nitrogen as nitrate.

A real field-testing method that can be used is using a pot test. Paddy, corn, and red onion are commonly used as a plant model to determine the release behavior of the SRF. Usually, this test uses randomized block design.

Fig. 6.9 shows a clear example of the water dissolution test of a zeolite-based SRF with different manufacturing methods. The SRF showed slow release properties for 100 days, while the urea represents the conventional fertilizer that dissolved in water for less than 2 h. The rate of release is controlled according to the binder. In addition, the release rate of SRF that utilizes water glass and molasses binder were quite similar while starch tends to prolong the nutrient dissolution. The rate of release in the first period of less than 2 days was 50% of total nutrients and gradually decreased until 100 days. The effect of binder on the release profile of matrix-based

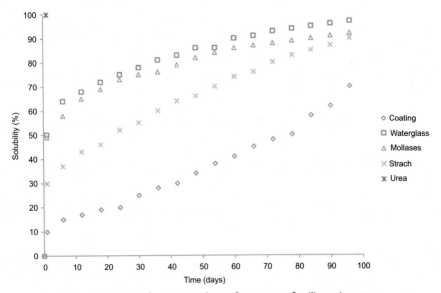

Figure 6.9 Rate of nutrient release from some fertilizers in water.

fertilizer has been documented [16,18]. Meanwhile, the CRF is superior compared to any SRFs in terms of reducing the dissolution rate of the encapsulated nutrient.

Fig. 6.10 shows a sample of the soil incubation test of ammonia and nitrate concentration. The availability of N in NH_4 increased on the third day with the peak value obtained on the 14th day before it decreased. The availability of N as NO_3 in soil kept increasing from the beginning, and this corresponded to the transformation of ammonia released from fertilizer into nitrate by the microorganism.

5. Conclusion

In conclusion, the demand for SRF/CRF is rapidly growing due to environmental consideration and also due to precision farming practices. Its various manufacturing methods are developed for achieving a specific target of nutrients while reducing the manufacturing cost, as the hindrance of CRF application is still at a higher price compared to conventional fertilizer. Coated fertilizer is released over a predictable nutrient, while matrix-based SRF is commercially used in a vast plantation as this type tends to blend essential nutrients targeted to a specific commodity with a single or few application times.

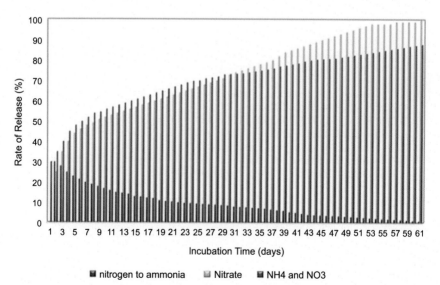

Figure 6.10 Relationship of incubation time and availability of nitrogen as NH_4 and nitrate in the soil.

References

[1] H. Aoki, Y. Sekiguti, Development and application of coated fertilizer in Japan, J. Food Sci. Eng. 9 (2019) 142−152, https://doi.org/10.17265/2159-5828/2019.04.004.

[2] B. Azeem, K. Kushaari, Z.B. Man, A. Basit, T.H. Thanh, Review on materials & methods to produce controlled release coated urea fertilizer, J. Control. Release 181 (2014) 11−21.

[3] X. Bai, Y. Wang, X. Huo, R. Salim, H. Bloch, H. Zhang, Assessing fertilizer use efficiency and its determinants for apple production in China, Ecol. Indicat. 104 (2019) 268−278.

[4] S.G. Belak, R.H. Campbell, Patent No. 3,242,237, U.S. Patent and Trademark Office, Washington, DC, 1966.

[5] J.H. Detrick, Patent No. 5,599,374, U.S. Patent and Trademark Office, Washington, DC, 1997.

[6] T. Fujita, Y. Yamashita, S. Yoshida, K. Yamahira, Patent No. 5,009,696, U.S. Patent and Trademark Office, Washington, DC, 1991.

[7] T. Fujita, Y. Yamashita, S. Yoshida, K. Yamahira, Patent No. 5,176,734, U.S. Patent and Trademark Office, Washington, DC, 1993.

[8] J.L. Havlin, S.L. Tisdale, W.L. Nelson, J.D. Beaton, Soil Fertility and Fertilizers: An Introduction to Nutrient Management, eighth ed., Pearson, New Jersey, 2014.

[9] Y. Kusumastuti, A. Istiani, Rochmadi, C.W. Purnomo, Chitosan-based polyion multilayer coating on NPK fertilizer as controlled released fertilizer, Adv. Mater. Sci. Eng. (2019), https://doi.org/10.1155/2019/2958021.

[10] B. Ni, M. Liu, S. Lü, Multifunctional slow-release urea fertilizer from ethylcellulose and superabsorbent coated formulations, Chem. Eng. J. 155 (2009) 892−898.

[11] J.M. Ogle, J.D. Sims, Patent No. US 8,741,022 B2, U.S. Patent and Trademark Office, Washington, DC, 2014.

[12] C.W. Purnomo, S. Indarti, C. Wulandari, H. Hinode, K. Nakasaki, Slow release fertiliser production from poultry manure, Chem. Eng. Trans. 56 (2017) 1531−1536, https://doi.org/10.3303/CET1756256.

[13] C.W. Purnomo, A. Respito, E.P. Sitanggang, P. Mulyono, Slow release fertilizer preparation from sugar cane industrial waste, Environ. Technol. Innov. 10 (2018) 275−280, https://doi.org/10.1016/j.eti.2018.02.010.

[14] A. Shaviv, R.L. Mikkelsen, Controlled-release fertilizers to increase efficiency of nutrient use and minimize environmental degradation − a review, Fert. Res. 35 (1993) 1−12.

[15] A.R. Shirley Jr., R.S. Meline, Patent No. 3,903,333, U.S. Patent and Trademark Office, Washington, DC, 1975.

[16] E.P.O. Sitanggang, C.W. Purnomo, The effects of binder on the release of nutrient from matrix-based slow release fertilizer, Mater. Sci. Forum 886 (2017) 138−144, https://doi.org/10.4028/www.scientific.net/MSF.886.138. MSF.

[17] Q. Taomei, L. Shaoyu, L. Tao, C. Jiao, H. Mengjie, J. Yanzheng, et al., A multielement compound fertilizer used polydopamine and sodium carboxymethyl starch matrices as coatings, Int. J. Biol. Macromol. 124 (2019) 582−590.

[18] H. Windia, W.P. Chandra, P. Suryo, Slow release npk fertilizer preparation from natural resources, Mater. Sci. Forum 948 (2019) 43−48, https://doi.org/10.4028/www.scientific.net/MSF.948.43.

CHAPTER 7

Controlling factors of slow or controlled-release fertilizers

**Ikram Ganetri[1], Youness Essamlali[1], Othmane Amadine[1],
Karim Danoun[1], Soumia Aboulhrouz[1], Mohamed Zahouily[1,2]**
[1]VARENA Center, MAScIR Foundation, Rabat Design, Rabat, Morocco; [2]Laboratoire de Matériaux,
Catalyse et Valorisation des Ressources Naturelles, Université Hassan II-Casablanca, Morocco

1. Introduction

Slow and controlled-release fertilizers (S-CRF) are granulated fertilizers that release nutrients gradually into the soil in order to synchronize the supply of nutriment with the requirements for model crop growth [1]. Controlled-release fertilizer (CRF) and slow-release fertilizer (SRF) are often simply, though inaccurately, employed as synonyms; nonetheless, they are slightly different. According to Trenkel and Shaviv [1,2], the term controlled-release fertilizer became acceptable when applied to fertilizers in which the factors dominating the rate, pattern, and duration of release are well known and controllable during CRF preparation. While slow-release fertilizers involve the release of the nutrient at a slower rate than usual by providing a transport barrier to the fast dissolution of nutrients in water when exposed without a coating [3], but the rate pattern and duration of release are not well controlled. In this chapter, a specific interest will be attributed to the CRFs due to the potential control of their release pattern in the soil. In fact, nutrient release profiles from CRFs are intended to be engineered to have optimal nutrient release, i.e., the rate of release of nutrients from the CRFs matches the maximum uptake rate of nutrients by plants, which varies dynamically throughout the development stages of the plant over the growing season [3]. This makes CRFs a good alternative to conventional fertilizers owing to their ability to reduce the wastage of capital that causes serious environmental hazard, affecting both land and water ecosystems, and to improve the nutrient use efficiency of plants by increasing their nutrient uptake, with enhanced yields in terms of quality and quantity [4].

Various delivery materials and techniques have been introduced to ensure the delay of the release of different chemical ingredients for specific

Controlled Release Fertilizers for Sustainable Agriculture
ISBN 978-0-12-819555-0
https://doi.org/10.1016/B978-0-12-819555-0.00007-8

111

applications [1,4]. Generally, there are three common methods of making S-CRF: by preparing less soluble chemical compounds, coating using relatively inert materials, and incorporating into matrices [5]. Nutrients release behavior and its mechanisms are mandatory to understand in order to design the appropriate CRF that retards fertilizer release to such a slow pace that a single application to the soil can meet nutrient requirements for model crop growth [1]. Several authors attempted to study experimentally and theoretically the release behavior of different kinds of CRF. In the present chapter, we will try to deliver to the researcher and industry a summarized overview of the influence of different factors (factors depending on the CRF type itself and factors depending on the external parameter [soil, temperature ...]) on the delivery profile of nutrients and their impact on the crops.

2. Fertilizer's composition and shape

2.1 S-CRF generalities and interest

The application of S-CRFs is a good approach for reducing nonpoint contamination in agriculture and acquiring higher nutrient use efficiency. Indeed, in order to keep world crop productivity in step with human population growth, the application of fertilizers has extensively increased, which has led to serious economic losses and environmental threats, such as water eutrophication and toxicity, groundwater pollution, air pollution, soil quality degradation, and even the ecosystem change [6,7]. To overcome these challenges, considerable efforts have been made to develop efficient and suitable methods for the preparation of highly nutrient-efficient S-CRFs, which have been inspired by drug delivery systems, to address the mismanagement of basic fertilizers in agriculture.

Two principal methods are commonly employed for the production of S-CRFs: (a) matrix method or/and (b) core-shell method.

(a) The matrix method consists of a chemical mixture of the fertilizer formulation and the matrix responsible for nutrient release delay;

(b) The core-shell method is achieved by applying a coating of protective layer of water-insoluble, semipermeable or impermeable-with-pores material. These kind of S-CRFs are mainly obtained by spray-drying method using either drum-coating (called also: pan-coating) or fluidized bed. The release profile depends on the nature of the coating layer.

2.2 Solubility and bioavailability of nutriment

Nitrogen (N), phosphorus (P), and potassium (K) are the major macronutrients commonly used in agriculture worldwide. Depending on their nature and origin, CRFs are generally classified into single-nutrient fertilizers such as polymer-coated urea (PCU) and multiple nutrient fertilizers containing three main nutrients: nitrogen, phosphorus, and potassium (N, P, and K), and many formulations include calcium, magnesium, sulfur, and micronutrients. The term availability (phytoavailability) refers to the readily soluble fraction of element that is taken up by the plant [8].

The availability of nutrient in the soil influences the release mechanism of the nutrient from fertilizers. According to several authors [1−3,9,10], macromolecular coated fertilizers are expected to undergo a release process called "diffusion mechanism" which can be described by the following steps (Fig. 7.1):

(1) Penetration of water inside of the polymer coating;

(2) Swelling by the absorbed water;

(3) Partial nutrient dissolution out of the granule fertilizer inside of the coating polymer;

(4) Slow-release of nutrients via diffusion, under concentration or pressure gradient or a combination thereof, out of the polymer matrix.

The diffusion rate of nutrient from the polymer coating to the soil depends on the solubility and availability of the nutrient in the soil itself. In fact, the gap of concentration between the two mediums (soil and the core of the CRF) governs the rate at which the nutrient is released. When the nutrients naturally present in the soil are not available (not readily soluble), the nutrient release from the CRF to the soil increases. The availability of N from organic or mineral sources varies considerably [11]. Plants take up

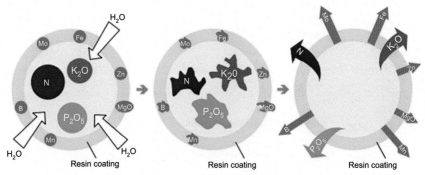

Figure 7.1 The fertilizer nutrient diffusion from coating to the soil [2].

N normally as nitrate or ammonium ions, through their roots from the soil solution [2]. However, the root system of most arable crops only explores 20%–25% of the available soil volume in any oneyear due to the competition between soil and plant roots for available nutrients in the soil–plant system [12]. Hence, a considerable proportion of applied fertilizer-N is lost during the year of application, by one of the three types of processes/reactions that decreases its availability to plants [2,13].

- Microbial: nitrification, denitrification, and immobilization;
- Chemical: exchange, fixation, precipitation, hydrolysis;
- Physical: leaching, run-off, and volatilization.

In order to overcome N deficit due to its loss and to increase the efficiency of nitrogen fertilizers, four major classes of controlled-release N fertilizers are proposed [14], (Fig. 7.2):

(i) Encapsulated soluble fertilizers (e.g., Prokote Plus, Osmocote, sulfur-coated urea);

(ii) Low-solubility inorganic salts (e.g., magnesium ammonium phosphate);

(iii) Low-solubility organic compounds which chemically and microbially decompose (e.g., oxamide, urea-formaldehyde, isobutylene diurea);

(iv) Water-soluble materials that gradually decompose (e.g., guanylurea).

The choice of the kind of N CRF to apply depends on several external parameters such as temperature, moisture, biological activity, and wetting and drying cycles. Some of these parameters will be discussed in the upcoming paragraphs.

Phosphorus is the second most supplied macronutrient to crops in order to increase the quantity and raise the quality of food production. Conventional mineral P fertilizers manufactured from phosphate rock might be easily leached in sandy soils, adsorbed, and ultimately fixed in acid and highly weathered soils [15]. Irreversible P fixation is favored by high

$$(NH_2)_2CO + 2H_2O \xrightarrow{\text{Urease}} (NH_2)CO \quad \text{(1)}$$

$$(NH_4)CO_3 + 2H^+ \xrightarrow{\text{Ammonification}} 2NH_4^+ + CO_2 + H_2O \quad \text{(2)}$$

$$2NH_4^+ + 3O_2 \xrightarrow{\text{Ammonification}} 2NO_2^- + 2H_2O + 4H^+ + Energy \quad \text{(3)}$$

$$2NO_2^- + O_2 \xrightarrow{\text{Nitrobacter bacterium/nitrification}} 2NO_3^- + Energy \quad \text{(4)}$$

$$NO_3^- \xrightarrow{\text{Microorganisms/O_2 deficient soil}} N_2 + N_2O \quad \text{(5)}$$

$$NH_4^+ \xrightarrow{\text{Urease enzyme/Basic soil pH}} NH_3(g) + H^+ \quad \text{(6)}$$

Figure 7.2 Urea transformation to produce plant-available nitrogen [1].

contents of (hydro)oxides of Fe and Al, hence reducing its availability to the crop and limiting its nutrient use efficiency [12]. Since the fixation of P in the soils depends strongly on its acidity, the impregnation of acid phosphate sources with MgO seems to be an efficient strategy to eliminate acidity and increase P use efficiency by plants [12,15].

If the problems related to N and P loss inland fields are well established, the K release profile is hard to be generalized because of the complex mechanism involving physical interaction and possible chemical bonding that might occur between K species and other minerals included in the composition of the CRFs. Moreover, the small size of potassium ions helps in easily entering micropores inside the carbon content of the matrix as well. The strong attraction of the matrix to the potassium ions by adsorption and charge attraction mechanism can be the origin of the sluggish release of this nutrient [5].

In the case of complex fertilizers where two or more nutrients are mixed, the understanding of the effect of the combination of several nutrients in one granule on nutrients release rate and pattern is of utmost importance. Although significant knowledge regarding nutrient release from single-nutrient fertilizer such as PCU was gained, less attention was devoted to the release of nutrients from compound NPK coated CRFs, in which the release process is qualified as more complex than with a single fertilizer/nutrient. According to Du et al. [16], the release of each nutrient dependents highly on its own solubility in solution, diffusivity/permeability through the polymer coating, and interactions between other nutrients, as well as the incubation conditions notably temperature, water content, and medium type. In general, nitrate release was the fastest, followed by ammonium and potassium whereas phosphate has a significantly slower release rate [16]. In their study, related to release characteristics of slow-release compound fertilizers, Wilson and Chem [16] demonstrated that the fractional rate of release of N was greater than that of K, and even more than the release rate of P. The main factor controlling the differential release of nutrients, at a given moisture level and temperature, was assumed to be the degree of solubility of the ions, with N being the most rapidly soluble and P being the least soluble [17].

2.3 Fertilizer's shape: granulate

The key feature of the fertilizer is that it must be granular. Industrially, only granulated fertilizers are suitable for coating treatment in order to produce controlled-release fertilizers. Granulation can have a large effect on the effectiveness of fertilizers, considering that powdered fertilizers are generally impractical for application and generate a high amount of dust during their storage and manipulation. Making loose powder fertilizer into granulate shape has several advantages. It provides better handling, reduces transportation and storage cost, slows down the release rate of the nutrients, helps to reduce segregation, and improves the content uniformity of the final product [12,18–20]. The surface characteristics of fertilizer granules are one of the key factors that significantly affect the release rate of S-CRFs. Irregular granules are more difficult to coat uniformly which can lead to the apparition of imperfections and defects in the coating layer thus resulting in an accelerated release pattern. The particle size distribution of the granules is also very important. Products with high uniformity and repeatable particle size distribution will produce a consistently better final product. Rotating drums are widely used to produce granulates or pellets in the fertilizer industry. This technology is also applied for coating purposes [18]. Granulate dissolution speed is determined by the granule size, its distribution, and porosity. The higher the granule porosity the smaller the relative volume to dissolve per granule, which induces a quicker the dissolution [18].

Among the main operational factors that affect the granulate shape are the granulation process parameters: the intensity of powdered raw material delivery, the inclination angle of the coating drum, rotational speed, and the residence time of the material. The disc axis is the major factor that affects the residence time of the material in the granulator. The augmentation of the inclination angle of the disc axis to horizontal position extends the residence time of the material, thus producing granulates with incremented size and greater homogeneity. The increase in the rotational speed of the disc raises the granule's size [21]. Fig. 7.3 illustrates the different shapes of fertilizer's granulate produced by granulation [18].

3. Coating composition and physical-chemical properties

3.1 Type of coating

The uptake of macronutrients by seasonal crops takes place generally in a "sigmoidal" way (S-shaped) as shown by the scheme in Fig. 7.4. Hence, the ideal fertilizer should release nutrients in a sigmoidal pattern for optimal plant nutrition and reduction in nutrient losses by processes that compete with the plant's nutrient requirements [2]. Coating the commercial fertilizer

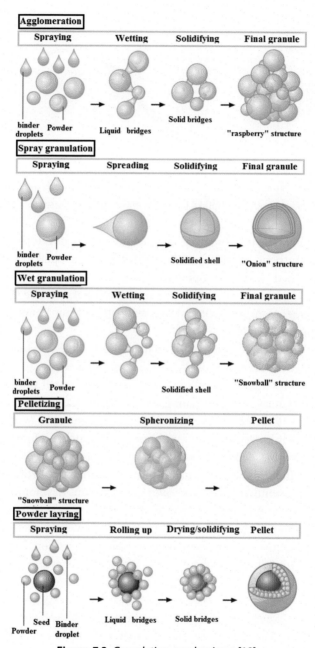

Figure 7.3 Granulation mechanisms [18].

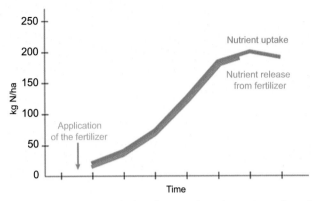

Figure 7.4 Profiles of nutrient uptake by plants and nutrient release from fertilizers [2].

pellets with any type of substances has been intensively developed aiming to match the nutrient supply profile with the crop uptake kinetic [5]. Fertilizers coated with hydrophobic materials provide reasonable/good control over the rate of release. The pattern of temporal release from coated fertilizers ranges from parabolic release (with or without "burst") to linear release to sigmoidal release.

In fact, fertilizers' coating with hydratable, soluble, or biodegradable polymers is an effective way of providing controlled release of nutrients to the soil [22]. This method allows high nutrient content per total weight by just adding a thin layer of the coating on the surface [23], while the matrix-based fertilizer will reduce the nutrient content per total weight of the fertilizer, since it has to be mixed with inert solids such as chars or clays by mostly 50% wt [5]. Hence, we will limit our discussion to polymer-coated CRFs due to their higher potential to provide nutrients to crops. Polymer-based coating formulations have known a tremendous development during the last decades in CRFs due to the wide amount of materials suitable for such application and the versatility of properties they may offer in order to meet the specific needs of each harvest, soil composition, and weather conditions. The following scheme (Fig. 7.5) summarizes the main type of polymers applied in fertilizers' coating in order to control the release profile of nutrients:

Numerous experimental studies have been reported on nutrient release from CRFs using a variety of coating materials, including conventional nondegradable polymers (polysulfone, polyolefins, polyethylene, poly-urethane, polystyrene, acrylic acid-co–acrylamide …), and biodegradable polymers (lignin, carboxymethyl cellulose, starch, alginate beads, and

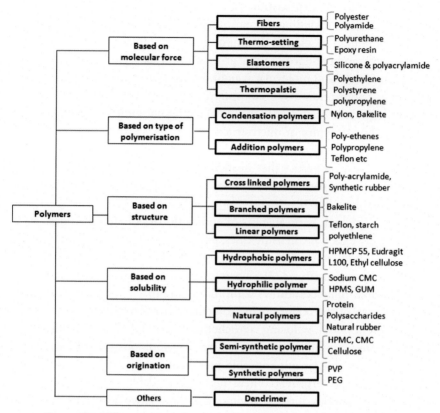

Figure 7.5 Different types of polymers employed as a coating for CRFs.

chitosan ...) [3,23]. The selection of the appropriate polymer matrix to use as a coating relies on the physical-chemical, chemical resistance, and mechanical properties of the polymer films itself. Water and/or water vapor permeability are inherent properties of polymers that may vary with the coating thickness and the amounts of imperfections in the coating layer. These properties dictate the rate at which the nutrients inside the coating layer are dissolved; hence, they control the release profile. In many cases, a desired amount of imperfections is deliberately introduced into the coating layer in order to meet the specific need of the plants. When it comes to mechanical properties, the elasticity or brittleness of the polymer affects the resistance of the coating layer to handling operations, which influences the quality and uniformity of the coating layer function of time [24]. Adding to the abovementioned characteristics, the degradation process is a key property of the polymer that controls the nutrient's liberation profile of

CRFs. The rate of degradation of a polymeric material is affected by numerous factors, including among others: the polymer structural features such as configuration, chemical composition, distribution of repeating units in multimers, and presence of ionic groups. Its physiological features such as molecular weight and molecular weight distribution, processing conditions, annealing, absorption or adsorption, and physiochemical factors. Morphological features include structural configuration (e.g., amorphous or semicrystalline structure or microstructures), shape, specific area, and site of implantation. The degradation of the polymer depends also on certain specific physical properties that strongly affect the rate of polymer degradation. The water permeability and solubility of polymers determine their rate of hydrolysis [9,25].

Despite the high performances of conventional polymer materials, they can be highly expensive in some cases, pollutants, or toxic and most difficult to degrade with potential environmental impacts (accumulation); these are the major concerns [22]. Therefore, the inclusion of fertilizers to biodegradable films constitutes a promising approach that could simultaneously lessen plastic pollution and soil eutrophication, as well as increasing crop efficiency and reducing agrochemicals use [9]. Natural polymeric carbohydrates appear as an alternative to nonbiodegradable materials acting as permeable or impermeable membranes with tiny pores in slow-controlled release fertilizers. These highly degradable materials have also received tremendous interest owing to their relative low cost and low environmental damage due to their biodegradability and low accumulation in the environment [23].

Although natural polymers have some obvious advantages over other coatings materials, these kind of polymers suffer from some drawbacks related to their inherent physicochemical and mechanical properties such as their poor film-forming ability and their sensitivity to water, which may limit their ability to act as a barrier to control the release rate of nutrients. Thus, it is necessary to modify the raw material in order to develop new green bio-based polymer coating materials with superior properties for the production of controlled-release fertilizers [5,23,27]. Natural polymer blends in various combinations with hydrophobic synthetic polymers have shown to be a promising strategy to enhance the liberation control while preserving the biodegradable character of the coating material. The introduction of inorganic nanoparticles can reduce the cost and improve the properties of the natural polymer coating [28]. Fertilizer release could induce in situ microorganisms growth that competes for the take up of the nutrients released, and meanwhile, grow on the surface of CRFs and start the events of depolymerization [26]. The blend of two polymers to form semi-interpenetrating polymer networks seems also an efficient method to overcome natural polymers limitation. The feature of this network is the penetration of some branched or linear macromolecules into a molecular dimension of networks; the product generated by this blending method usually displays exceptional performances superior to either single natural polymer [29]. Superabsorbent hydrogels and hydrogels are also very attractive to use in agriculture as a coating material for fertilizers. These materials are kind of hydrophilic polymers with three-dimensionally cross-linked structure that can absorb a large amount of water and saline solutions due to the presence of hydrophilic functional groups in their unique three-dimensional network structure [29,30]. Hence, they could reduce the consumption of irrigation water and improve the amount of fertilizer retained in soil [28]. Majeed et al. [26] did a literature survey and noticed that overall polymer-coated CRF's, six main sources of biodegradable polymeric materials are used under the current research and development:

- CRFs coated with pure/modified natural polymers;
- CRFs coated with mixed natural polymers;
- CRFs coated with nature-inspired/bio-derived synthetic polymers;
- CRFs coated with mixed natural-bio-derived synthetic polymers;
- CRFs coated with mixed natural-hydrocarbon-derived synthetic polymers;
- CRFs coated with hydrocarbon-derived synthetic biodegradable polymers.

The understanding of the impact of microbial susceptibility, type, and degree of cross-linking; porosity; and polymer composites' composition are few of the coating material properties that could help to engineer the desired biodegradation in the coatings of fertilizers to gain better control over current problems of the nutrient release with these materials [26].

3.2 Coating layer properties (porosity, thickness, ...)

At industrial scale, the coating of granular fertilizers by polymer solutions is commonly achieved by spraying method in pan-discs or rotary drum. The quality of the coating layer is governed by the solution percentage, its temperature of application, the pan-disc rotation speed, inclination angle, drying temperature, and used solvent. In fact, the CRF preparation parameters are important to determine the nutrient's release profile. It is shown that each nutrient has a different release behavior even with the same matrix by different preparation parameters [5].

The release rate and mechanism depend intimately on the porosity and the thickness of the polymer coating layer. The coating layer thickness is controlled by varying the amount of coating solution sprayed and the number of layers, while the porosity is a unique property of a polymer and its process to coat the CRFs [26]. In the case of hydrogels, the pore size and porosity could be tuned by changing various parameters, such as concentration of polymers or monomers, cross-linking degree, temperature, freezing rate, addition of organic solvents or inorganic particles. Hydrogels with lower porosity or larger pore sizes degraded more rapidly than those with higher porosity or smaller pore sizes [31].

4. Soil parameters

The release rate and profile depend strongly on soil properties. After soil contact, the biodegradation of the coating polymer and the release of nutrients from CRF are controlled by a synergistic triad. Many physical (pH, temperature, moisture content, wetting, and drying), chemical (soil composition), and biological (microbial activity) factors of the soil govern the final release mechanism of nutrients [26,32].

4.1 Soil biological/microbial activity

The disintegration of the coating is required for the release of about 15%–30% of the fertilizer that remained unreleased from polymer-coated CRFs due to the concentration gradient difference across the polymer coatings. Biodegradation of CRFs is a complex process that involves cascade events controlled by abiotic and biotic stimuli in the environment [26]. The degradation process of biodegradable polymer coatings is described by Versino el al. [9] as follows: The microorganisms colonize the surface of the polymers, secreting enzymes responsible for the hydrolysis of the polymeric bonds generating oligomers, dimers that are disintegrated into monomers. These monomers are then converted to carbon dioxide, water, minerals, and biomass under aerobic conditions, or to carbon dioxide, methane, and humic material in the case of anaerobic biodegradation without leaving any potentially harmful substances [9,33].

Soil's biological/microbial activity itself depends on the soil nature, its temperature, moisture, pH, and other environmental conditions [34]. Many studies have investigated the influence of soil type on the release profile of S-CRFs. There are several types of soils but the most common ones are silt, clay, sandy, and loamy soils. The silt is a smooth type of soil with particle size >2 μm. It is a rich and fertile soil since it retains water and nutrients. Clay soil is a very compact type of soil with particle size <2 μm. The small particles tend to pack closely together which allows it to retain a lot of water and nutrients. These soils are known to be very rich and very dense in nature. Sandy soils are characterized by less than 18% clay and more than 68% sand in the first 100 cm of the sol [35]. Loam is soil composed mostly of sand (particle size > 63 μm), silt, and a smaller amount of clay. By weight, the mineral composition is about 40%−40%−20% concentration of sand−silt−clay, respectively [36]. In general, loam soils contain more nutrients, moisture, and humus than sandy soils; have better drainage; and infiltration of water and air. On the contrary, sandy soils are described as soils with a weak structure or no structure, poor water retention properties, high permeability, high sensitivity to compaction [35]. In sandy soils, the bigger the holes are, the higher the oxygen circulation is, which enhances the activities of microorganisms responsible for the biodegradation of polymer coating [26]. In general, the higher moisture content in soils and higher temperatures increase the microbial activity, leading to the dominance of communities that have the ability to access or metabolize substrates more readily [32]. Moreover, in loamy soil the nitrification rate of Ammonia is faster than in sandy soil, which contributed to less availability of NH_4^+, thus, less NH_3 volatilization in loamy soil versus enhanced NH_3 volatilization in sandy soil [37].

The data of some soil microflora responsible for the degradation of some types of coatings are summarized in the following table (Table 7.1):

4.2 Soil physical-chemical properties

The degradation of polymer coating depends strongly on their physical-chemical properties; hence, the release behavior of CRFs coated by polymers depends also on those parameters. In fact, many authors have studied the influence of soil temperature, pH, and water content.

Qi et al. [40] have developed a novel adsorption material using carboxymethyl cellulose to hold water and release fertilizer slowly in saline-alkaline soils. The authors studied the influence of external pH on the swelling behavior of the carboxymethyl cellulose immobilized controlled-release fertilizer microspheres. According to the same study, as the pH value increased from 2 to 6, the adsorption capacity increased gradually, while a further increase in pH from 8 to 12 caused a steep decrease in the adsorption capacity (Fig. 7.6). The higher swelling capacity at low pH values was explained by the strong hydrogen bonding interactions among

Table 7.1 Enzymes or bacteria responsible for the degradation of some biodegradable polymer used as coating materials.

Polymer coating material	Characteristics	Biodegradability	Enzymes or bacteria responsible of the biodegradability
Polyurethane	Synthetic polymer	Partially biodegradable under microbialaction	Bacteria that secrete "Polyurethanase" enzyme such as *Petalotiopsis micropora*
Chitosan	Polysaccharide derived from the alkaline deacetylation of chitin, abundant and nontoxic	Degrades completely	Bacteria producing "chitosanases" : *Streptomyces* spp. *Kitasatospora* spp
Alginate	Natural polymer and nontoxic	Degrades completely	Alginate hydrolase enzyme: *Asteromyces cruciatus* and *Dendryphiella salina*
Starch	Natural polymer easily modified and nontoxic [38]	Degrades completely	Amylases, amyloglucosidases, and transglucosidases [39]
Lignin	Readily from plant derivatives, low cost	Degrades completely	Decomposes under the Laccas Produced by either the Actinomycetes, α–Proteobacteria, γ–Proteobacteria

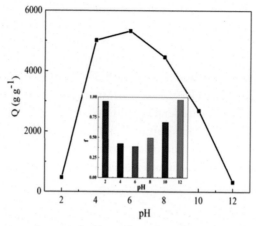

Figure 7.6 Influence of external pH on the adsorption capacity of carboxymethyl cellulose immobilized controlled-release fertilizer microspheres [40].

the $-COOH$ groups and the physical cross-linking reaction would occur more easily. In addition, studies have shown that electrostatic repulsion among the $-COO^-$ group decreases and the 3D network shrinks at low pH value. As pH values increased, the number of $-COO^-$ groups in the polymer chain would have increased. This would have destroyed the hydrogen bonds and increased electrostatic repulsion among the $-COO^-$ groups.

In another study, Araújo et al. [22] have evaluated the behavior of chitosan coating during the release of urea under different pH (2.5, 4.0, and 9.0). The authors studied three types of CRFs: chitosan–humic acids–urea, chitosan–peat–urea, and chitosan–humin–urea. At pH 2.5, due to the solubility of chitosan in acidic medium, the release of urea was facilitated while the presence of humic acid or peat would not affect the release of urea in opposite the aliphatic compounds lowered the affinity between urea and solvent, permitting a better control of urea release. At pH 4.0, the release of urea from chitosan coating containing humic acid was more controllable. This behavior was explained by the ionization of the $-$ COOH groups of the humic substances at pH 4.0. Consequently, the diffusion of water in the polymeric medium enabled the solubilization of urea, which was then available for release into the aqueous environment. However, electrostatic interaction of urea with the COO^- groups would introduce an additional step in the release process, resulting in slower urea release. At pH 9.0, the release rate was the slowest for urea CRF-coated with chitosan.

Most of the authors agree that soil temperature increase implies an increase in the N release [32,37]. Chen and Chen [41] have studied the temperature responsiveness of a complex fertilizer (N—P—K: 20-20-20) coated by methylcellulose or hydroxypropyl methylcellulose hydrogels. The release experiments under 25°C and 35°C have proven the influence of temperature on the release rate of nutrients. In fact, release profiles of the prepared CRFs considerably differed between 25°C and 35°C. The temperature-responsive fertilizer release can be attributed to the temperature-sensitive segment of the methylcellulose and hydroxypropyl methylcellulose and the Hofmeister effect of the hydrogels.

5. Conclusion

Extensive efforts to solve fertilizers losses have brought about a variety of strategies. In particular, slow-controlled release fertilizers (S-CRF) provide an effective way to improve nutrient-use efficiency, minimize nutrients losses by different mechanisms. A series of studies have been carried out to develop coating materials to ensure controlled release of fertilizers. Several parameters must be controlled and monitored in order to improve S-CRFs' understanding of the nutrient release mechanism.

In this chapter, we have discussed different factors influencing the liberation profile of the CRFs. The release of fertilizer takes place via a complex mechanism due to the interactions of several factors: the size and the shape of the fertilizer, the type of the coating layer and its quality, as well as the properties of the soil. The S-CRF should address the nutrient needs of crops taking into account the properties of the soils. A better understanding of the effect of temperature, humidity, and soil pH and soil bioactivity would provide new opportunities for more efficient coatings.

Different types of polymers used as fertilizer coatings have been cited in the literature, including biopolymers. Being cheap, biodegradable, and environmental friendly, polymers derived from the natural resources are considered as very interesting substitutes to regular polymers. However, their augmented hydrophobicity and poor water resistance are the main limitations that restrict their wide use as fertilizer coating. The combination of natural polymers with hydrophobic synthetic polymers or inorganic nanoparticles or superabsorbent polymers has been proven to be a promising strategy to improve the control of fertilizer release while maintaining the biodegradability of the coating material.

Adding to the development of the S-CRF, the promotion of modern and sustainable agriculture has also been associated with the progress in the sustainability of water and pesticide use, energy inputs, manufacturing, and other economic sectors that also have a significant impact on the environment.

References

[1] B. Azeem, K. KuShaari, Z.B. Man, A. Basit, T.H. Thanh, Review on materials & methods to produce controlled release coated urea fertilizer, J. Contr. Release 181 (2014) 11—21.

[2] M.E. Trenkel, Slow-and Controlled-Release and Stabilized Fertilizers, International Fertilizer Industry Association (IFA), Paris, 2010.

[3] S.A. Irfan, R. Razali, K. KuShaari, N. Mansor, B. Azeem, A.N. Ford Versypt, A review of mathematical modeling and simulation of controlled-release fertilizers, J. Contr. Release 271 (Feb. 2018) 45—54.

[4] T.A. Wani, F.A. Masoodi, W.N. Baba, M. Ahmad, N. Rahmanian, S.M. Jafari, Nanoencapsulation of agrochemicals, fertilizers, and pesticides for improved plant production, in: Advances in Phytonanotechnology, Elsevier, 2019, pp. 279—298.

[5] C.W. Purnomo, A. Respito, E.P. Sitanggang, P. Mulyono, Slow release fertilizer preparation from sugar cane industrial waste, Environ. Technol. Innov. 10 (May 2018) 275—280.

[6] J. Chen, et al., Environmentally friendly fertilizers: a review of materials used and their effects on the environment, Sci. Total Environ. 613—614 (Feb. 1, 2018) 829—839. Elsevier B.V.

[7] A.M. Dave, M.H. Mehta, T.M. Aminabhavi, A.R. Kulkarni, K.S. Soppimath, A review on controlled release of nitrogen fertilizers through polymeric membrane devices, Polym. Plast. Technol. Eng. 38 (4) (Sep. 1999) 675—711.

[8] E. Brännvall, M. Nilsson, R. Sjöblom, N. Skoglund, J. Kumpiene, Effect of residue combinations on plant uptake of nutrients and potentially toxic elements, J. Environ. Manag. 132 (2014) 287—295.

[9] F. Versino, M. Urriza, M.A. García, Eco-compatible cassava starch films for fertilizer controlled-release, Int. J. Biol. Macromol. 134 (Aug. 2019) 302—307.

[10] G.Z. Zhao, Y.Q. Liu, Y. Tian, Y.Y. Sun, Y. Cao, Preparation and properties of macromelecular slow-release fertilizer containing nitrogen, phosphorus and potassium, J. Polym. Res. 17 (1) (Jan. 2010) 119—125.

[11] S. Agehara, D.D. Warncke, Soil moisture and temperature effects on nitrogen release from organic nitrogen sources, Soil Sci. Soc. Am. J. 69 (6) (Nov. 2005) 1844—1855.

[12] J.F. Lustosa Filho, C.F. Barbosa, J.S. da Silva. Carneiro, L.C.A. Melo, Diffusion and phosphorus solubility of biochar-based fertilizer: visualization, chemical assessment and availability to plants, Soil Tillage Res. 194 (Nov. 2019).

[13] A. Shaviv, Advances in controlled-release fertilizers, Adv. Agron. 71 (Jan. 2001) 1—49.

[14] R.L. Mikkelsen, H.M. Williams, A.D. Behel, Nitrogen leaching and plant uptake from controlled-release fertilizers, Fert. Res. 37 (1) (Feb. 1994) 43—50.

[15] J.F. Lustosa Filho, E.S. Penido, P.P. Castro, C.A. Silva, L.C.A. Melo, Co-pyrolysis of poultry litter and phosphate and magnesium generates alternative slow-release fertilizer suitable for tropical soils, ACS Sustain. Chem. Eng. 5 (10) (Oct. 2017) 9043—9052.

[16] C. Du, J. Zhou, A. Shaviv, Release characteristics of nutrients from polymer-coated compound controlled release fertilizers, J. Polym. Environ. 14 (3) (Nov. 2006) 223–230.

[17] S. Shoji, Controlled Release Fertilizers With Polyolefin Resin coating; Development, Properties and Utilization, 1992.

[18] J.P.K.S.A.D. Salman, M.J. Handslow (Eds.), Handbook of powder technology-Granulation, 2006.

[19] E.M. Hansuld, L. Briens, A review of monitoring methods for pharmaceutical wet granulation, International Journal of Pharmaceutics 472 (1–2) (2014) 192–201. Elsevier.

[20] Z. Zhang, et al., Experiments and modelling of potassium release behavior from tablet biomass ash for better recycling of ash as eco-friendly fertilizer, J. Clean. Prod. (2018).

[21] (PDF) A review on techniques for sago pearl granulation and sizing process [Online]. Available: https://www.researchgate.net/publication/276417956_A_Review_on_Techniques_for_Sago_Pearl_Granulation_and_Sizing_Process. (Accessed 22 October 2019).

[22] B.R. Araújo, L.P.C. Romão, M.E. Doumer, A.S. Mangrich, Evaluation of the interactions between chitosan and humics in media for the controlled release of nitrogen fertilizer, J. Environ. Manag. 190 (Apr. 2017) 122–131.

[23] M. Calabi-Floody, et al., Smart fertilizers as a strategy for sustainable agriculture, in: Advances in Agronomy, vol. 147, Academic Press Inc., 2018, pp. 119–157.

[24] H.M. Goertz, Technology developments in coated fertilizers, in: International Workshop on Controlled/Slow Release Fertilizers, Technion, 1993.

[25] M. Ul-Islam, S. Khan, M.W. Ullah, J.K. Park, Structure, chemistry and pharmaceutical applications of biodegradable polymers, in: Handbook of Polymers for Pharmaceutical Technologies, vol. 3, wiley, 2015, pp. 517–540.

[26] Z. Majeed, N.K. Ramli, N. Mansor, Z. Man, A comprehensive review on biodegradable polymers and their blends used in controlled-release fertilizer processes, Rev. Chem. Eng. 31 (1) (Jan. 2015) 69–95.

[27] Y. Li, et al., Synthesis and performance of bio-based epoxy coated urea as controlled release fertilizer, Prog. Org. Coating 119 (Jun. 2018) 50–56.

[28] H. Wei, H. Wang, H. Chu, J. Li, Preparation and characterization of slow-release and water-retention fertilizer based on starch and halloysite, Int. J. Biol. Macromol. 133 (Jul. 2019) 1210–1218.

[29] X. Li, Q. Li, X. Xu, Y. Su, Q. Yue, B. Gao, Characterization, swelling and slow-release properties of a new controlled release fertilizer based on wheat straw cellulose hydrogel, J. Taiwan Inst. Chem. Eng. 60 (Mar. 2016) 564–572.

[30] A. Olad, H. Zebhi, D. Salari, A. Mirmohseni, A. Reyhani Tabar, Slow-release NPK fertilizer encapsulated by carboxymethyl cellulose-based nanocomposite with the function of water retention in soil, Mater. Sci. Eng. C 90 (Sep. 2018) 333–340.

[31] E.S. Dragan, M.V. Dinu, "Advances in porous chitosan-based composite hydrogels: synthesis and applications, React. Funct. Polym. (Oct. 2019) 104372.

[32] X.H. Fan, Y.C. Li, Nitrogen release from slow-release fertilizers as affected by soil type and temperature, Soil Sci. Soc. Am. J. 74 (5) (Sep. 2010) 1635–1641.

[33] I. Kyrikou, D. Briassoulis, Biodegradation of agricultural plastic films: a critical review, J. Polym. Environ. 15 (2) (Apr. 2007) 125–150.

[34] P. Nardi, et al., Nitrogen release from slow-release fertilizers in soils with different microbial activities, Pedosphere 28 (2) (Apr. 2018) 332–340.

[35] A.C. Hartmann and G.L. Bruand, Physical Properties of Tropical Sandy Soils, n.d. http://www.fao.org/3/AG125E15.htm.

[36] https://en.wikipedia.org/wiki/Loam.

[37] X.H. Fan, Y.C. Li, A.K. Alva, Effects of temperature and soil type on ammonia volatilization from slow-release nitrogen fertilizers, Commun. Soil Sci. Plant Anal. 42 (10) (Jan. 2011) 1111−1122.

[38] K. Koch, Starch-based films, in: Starch in Food: Structure, Function and Applications, second ed., Elsevier Inc., 2017, pp. 747−767.

[39] P. Bernfeld, Enzymes of starch degradation and synthesis, in: Advances in Enzymology and Related Subjects of Biochemistry vol. 12, wiley, 2006, pp. 379−428.

[40] H. Qi, et al., Novel low-cost carboxymethyl cellulose microspheres with excellent fertilizer absorbency and release behavior for saline-alkali soil, Int. J. Biol. Macromol. 131 (Jun. 2019) 412−419.

[41] Y.C. Chen, Y.H. Chen, Thermo and pH-responsive methylcellulose and hydroxypropyl hydrogels containing K_2SO_4 for water retention and a controlled-release water-soluble fertilizer, Sci. Total Environ. 655 (Mar. 2019) 958−967.

CHAPTER 8

Sensors detecting controlled fertilizer release

Muhammad Yasin Naz, Shazia Shukrullah, Abdul Ghaffar
Department of Physics, University of Agriculture, Faisalabad, Punjab, Pakistan

1. Introduction

Currently, the world is facing the problem of an exponential increase in population. The population of this planet was around three billion in 1960, which is expected to reach nine billion in 2040 [1]. The world does not have too much time to ensure that there is enough land, food, water, and energy for the fast-growing population. The United Nations warns that if humans remain unsuccessful in curbing overpopulation, more than three billion people will be in poverty. However, the growing world food crisis presents an opportunity for researchers and investors in the farming sector, especially in fertilizer production. An increase in food demand brings a positive surge in fertilizer demand. As revealed by the International Fertilizer Association, the world uses 170 million tons of fertilizer every year for food, fuel, fiber, and feed. Among this, the nitrogenous fertilizers are being used to meet 48% of the total food demand of the world [2]. High fertilizer inputs augment the pollutants' level in soil, air, and water. The excessively available nitrogen during fertilization of crops also contributes to environmental pollution. The unassimilated reactive nitrogen acts as a pollutant and harms natural resources.

Most of the pollution in the world today is caused by human beings. Sources of manmade pollution include excessive fertilization of crops, sewage, stormwater, industry, automobiles, and burning of wood and fuels [3]. The contribution of nitrogen to the environment pollution is increasing with the growth in the human population. Owing to the large input of mineral fertilizers, the global production of crop and livestock has increased significantly over the past century. In response, the loss of reactive nitrogen to the air, water, and soil has augmented as well. The nitrogen cycle in terms of fixation, ammonification, nitrification, and denitrification is explained in Fig. 8.1. The major part of the nitrogen primarily comes from the industrial nitrogen fixation, which is directly contributing to the

Controlled Release Fertilizers for Sustainable Agriculture
ISBN 978-0-12-819555-0
https://doi.org/10.1016/B978-0-12-819555-0.00008-X

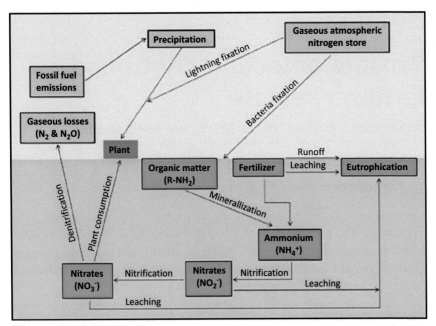

Figure 8.1 Nitrogen cycle explained in terms of fixation, ammonification, nitrification, and denitrification.

availability of nitrogen in terrestrial ecosystems. This increase in nitrogen in the environment is also responsible for the high deposition of nitrogen in agriculture-dominated landscapes [1,4]. The three largest sources of reactive nitrogen are nitrogen fertilizers, nitrogen in foods, and airborne nitrogen emissions. The food products account for the largest amount of reactive nitrogen (38%−75%) in the world. The airborne emissions of ammonia and nitrogen oxides and the subsequent deposition from the atmosphere contribute about 11%−36%, while nitrogen fertilizers contribute about 11%−32% to the reactive nitrogen [5]. The excess reactive nitrogen in the environment can lead to pollution problems, including poor air quality, acidification of lakes and rivers, disruption of foresting process, and degradation of coastal water quality.

Therefore, researchers are investigating different methods and mechanisms to minimize losses during fertilization. The fertilizer use efficiency can be improved in many ways, including nitrification inhibitations, urease inhibitation, and controlled release fertilizers. The use of controlled release fertilizers for slowing down the nutrients' leaching has recently been practiced by farmers worldwide. The added advantage of such fertilizers is the availability of controlled amount of nutrients in the soil for longer periods of time. The acceptable properties for fertilizer to qualify as a controlled release are different for different researchers. Under ambient conditions and absence of any external stress, Trenkel criteria for slow release fertilizers is (1) no more than 15% release in 24 h, (2) no more than 75% release in 28 days, and (3) at least about 75% release at the stated release time [6]. According to Tolescu and Iovu, controlled release fertilizers are those that contain at least one nutrient that either: (1) delays its availability in the up taking and utilization processes of the plants after application, or (2) is available for the plant in a significantly longer period than a standard considered to be a "quickly available fertilizer" [7]. According to Shaviv, the term controlled release fertilizers becomes acceptable when applied to fertilizers in which the factors dominating the rate, pattern, and duration of release are well known and controllable during the fertilizer preparation process [8]. This process involves the release of the nutrients at a slower rate than the usual rate; however, the rate, pattern, and duration of the release are not well controlled [9].

The release rate normally depends upon the nature and physical properties of the coating materials. The common materials, reported in the past literature, are neem (*Azadirachta indica* L.), resins, sulfur, natural carbohydrate polymers, and synthetic polymers [10,11]. A coating on granular fertilizer acts as a semipermeable or impermeable membrane with tiny pores and temporarily isolate the granules from the soil environment [12]. The release of nutrients through these membrane barriers depends on the physicochemical properties of the coating materials [13]. It is not significantly influenced by the properties of the soil such as salinity, pH, microbial activity, soil texture, and cation exchange capacity or redox potential. It reveals that monitoring of the nutrients' release from a coated fertilizer is not a trivial case. Many research efforts are underway to fully understand the release time, release rate, and mechanism of interaction of coating with water in the soil.

Different types of sensors are being developed to monitor the discharge parameters of controlled release fertilizers [14−17]. The word "sense" applies to the five senses of humans, and "making sense" defines our attempts to comprehend and understand the knowledge that might seem ambiguous or contradictory. For most plant farmers, field scouting and visual crop assessments are routine and provide data on crop growth, nutrient status, pest, weed, and the threat of disease. This information is used to prescribe management practices, such as irrigation water applications and herbicide treatments. The stress due to deficiency of nutrients, water shortage, or plant disease or pest infestation can be easily identified with the naked eye. However, plant stress issues that are more subtle or just developing may not be as obvious and may go unnoticed or misinterpreted. This chapter reviews the published literature on sensors for monitoring the performance parameters of controlled release fertilizers. Special emphasis is placed on nanosensors, optical sensors, electrochemical sensors, proximal sensors, airborne sensors, and remote sensing.

2. Nanosensors

Nanotechnology is at its primary stage in many areas, but seeing all the innovations, it indicates that it has a broad scope. There are always some challenges and rejections for any new technology, and nanotechnology is hitting new heights in its way to conquer both misconceptions and ethics. One recent avenue is the use of nanotechnology for developing nanosensors [15]. Nanosensors are nanoscale instruments that monitor physical quantities and translate them into signals, which can be interpreted and analyzed. Nanomaterial-based sensors have many advantages over sensors made from traditional materials in terms of sensitivity and specificity. Nanosensors may have enhanced sensitivity as they act on a scale comparable to natural biological processes, allowing functionalization with chemical and biological molecules, with events of recognition causing detectable physical changes. Compared to traditional methods of measurement such as chromatography and spectroscopy, nanosensors provide real-time monitoring during high-throughput applications.

2.1 Carbon nanotube-based sensors

Graphite and diamond have long been known as the only allotropes of crystalline carbon and their atomic structures were determined soon after the development of the X-ray diffraction method [18]. Carbon nanotubes (CNTs) were first discovered in 1991 by the Japanese scientist Sumio Iijima when two graphite electrodes were being dc arc-discharged for the production of fine particles and fullerenes. Since their discovery, CNTs have been extensively researched for an array of promising physical and chemical properties, such as high mechanical strength, electrical conductivity, chemical stability, and specific surface area [19]. Based on their structures, CNTs are classified into single-walled and multiwalled nanotubes. Fig. 8.2 illustrates commonly known structures of CNTs.

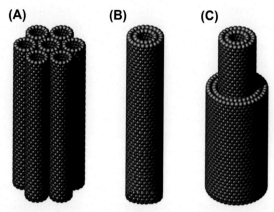

Figure 8.2 Illustration of (A) single-walled CNTs, (B) double-walled CNTs, and (C) multiwalled CNTs.

However, some other rare forms of CNTs, such as torus, fullerite, and nanoknot, are also possible [20]. Since the revelation of CNTs, tremendous experimental work has been conducted on them throughout the world on account of their high mechanical strength, current–carrying capacity, and other potential applications in the field of medical sciences, electronics, construction, aerospace engineering, coatings, gas adsorption, etc. [21,22]. Because of an array of practical applications, CNTs have been regarded as one of the best contenders for future applications in nanotechnology and their use in sensor devices is the most promising application in electronics. CNTs are being used as a sensing material in pressure, thermal, flow, gas, mass, optical, position, strain, stress, chemical, and biological sensors. When it comes to the agriculture sector, CNTs can be used to monitor humidity, temperature, gas concentration, plant growth rate, water intake, and uptake of nutrients from the soil.

Liu et al. [23] reported the effect of carbon-based materials combined with fertilizers on the crop growth parameters. The objective was to improve the crop yield and quality and to save the nitrogenous fertilizer for winter wheat. Qian and coworkers investigated the effect of carbon-based nanomaterial added fertilizers on the late rice in the double rice season area of Southern China [24]. It was reported that carbon added fertilizers improve the soil fertility, number of glume flowers per year, and rice yield. At the same time, the carbon added fertilizers were found to reduce the fertilizer release rate, thereby reducing the amount of fertilizer used and improving the efficiency of nitrogen uptake. The role of carbon nanomaterials in fertilizers to enhance plant growth and yield was well accepted with this positive result. These findings are vital in improving the efficiency of nitrogenous fertilizers, particularly the urea fertilizer for paddy growth. Nitrogen plays a critical role in the growth and productivity of paddy as it is necessary to synthesize many essential molecules, including nucleic acids.

Yatim et al. [25] study the role of pristine and functionalized multiwalled CNTs (MWCNTs) in improving the efficacy of urea fertilizer. Both types of MWCNTs were grafted onto urea fertilizer to produce urea-MWCNTs composite fertilizer. The growth of plants, treated with CNTs added urea fertilizer, was increased by up to 38.5%. The functionalized MWCNTs showed a more pronounced effect on plant growth. Adjizian et al. [26] developed CNT-based sensors for the detection of NO_2, CO, C_2H_4, and H_2O. The sensors were developed with CNTs doped with nitrogen and boron. These sensors were used for detecting the gases and water at ambient temperature and at 150°C. The nitrogen-doped CNTs were more sensitive to NO_2 and CO_2 while boron-doped CNTs showed high sensitivity to C_2H_4. The CNT-based sensors were highly sensitive to a change in humidity. Zaporotskova et al. [27] reported monolayer CNT-based sensors having vacancy surface defects. The defect containing sensors revealed high sensitivity to NO_2, NH_3, and H_2 as compared to defect-free CNT sensors.

In addition to nitrogen, plants also use carbon dioxide during photosynthesis. During photosynthesis, carbon dioxide combines with water in the presence of light to form sugar. Some part of the sugar converts into complex compounds necessarily required for the continuous growth of plants before reaching final maturity. However, if the level of carbon dioxide is too high, plants may not grow properly because carbon dioxide dissolves in water to produce carbonic acid. This acid makes the environment and soil more acidic and unsuitable for better plant growth. The use of CNTs in sensors for monitoring carbon dioxide can help control the concentration of carbon dioxide within the greenhouse and achieving an optimal environment for healthy plant growth. CNT-based pressure sensors can be used to spray liquid fertilizers, insecticides, and herbicides uniformly. A spray unit is normally operated by a tractor, consisting of a tank, pump, associated valves, and a series of spaced nozzles. The pressure sensor is used to sense nozzle pressure. A microprocessor monitors fluid pressure to ensure the distribution of the correct amount of liquid according to the tractor's speed.

2.2 Nanoaptamers

Aptamers are fascinating and promising biomaterials having high affinity and specificity against many valuable targets [15]. Nanostructures, combined with aptamers, are smart vehicles with extraordinary drug delivery properties. Aptamers are single-stranded nucleic acid that matches the target in all the ways by forming a three-dimensional structure with tight bonding with the target of interest. The selection of an appropriate aptamer has certain criteria. The systematic evolution of ligands by exponential enrichment sensors are more effective for detecting plant diseases, crop resistance to diseases, and crop yield [28]. The timely detection helps to eradicate the issues at their early stages. The insulin binding aptamers are used to monitor the extinction of light from cells. The toxicity in foods, caused by the herbicides and pesticides, can also be monitored by using nanoaptamers. When nanoaptamers are used for the controlled release of fertilizer, an aptamer is intended to link the chemical interactions between the root system and the soil microorganisms. Many polymer-coated controlled release fertilizers have been reported in the published literature [29,30]. The past reports show that in response to the aptamer's binding with the target, a polymer becomes more permeable. The polymer becomes enough permeable to deliver a payload of nutrients or medication.

The major challenge of an aptamer is to accurately identify specific chemical signals between the soil microbes and the rhizosphere of plants. Misidentification can lead to no release of nutrients or a suboptimal release rate. Can an aptamer still attach his target in a nanoscale film when immobilized? If not, the aptamer can misread the chemical signal, resulting in too much or too little delivery of nutrients or even no delivery at all. If a polymer is engineered as a permeable, it becomes less permeable or even impermeable due to the effect of target binding on the polymer's properties. As a result, polymer simply breaks down by unintelligently releasing the nutrients. In addition, an aptamer misreads the chemical signals and binds with the target incompletely if impurities are present in the payload of the nutrients. In this scenario, there is an incomplete or partial nutrient delivery.

Aptamers have close, if not superior, binding affinities to those of monoclonal antibodies [31]. The polyelectrolyte microcapsules containing aptamers in their walls that are specific for plant signals may also be used to monitor the release process. This will only cause the nutrient molecules to be supplied from inside the microcapsules when the plants need them [32]. Root exudate related aptamers are produced by utilizing systemic ligand evolution through exponential enrichment from a random DNA pool as well as from already existing aptamer pool. Such aptamers can work in the creation of an intelligent fertilizer network as molecular recognition probes. Researchers are near to reaching a nutrient breakthrough with huge ramifications for global agriculture [32].

The aptamers, synthetic DNA strands that differentiate chemical signals, were used by DeRosa and coworkers to detect a signal coming from the plant roots. When an aptamer obtains a root signal, it causes the coating to become unexpectedly more porous and release nitrogen. One of many problems, associated with aptamers, is to accurately identify specific chemical signals between the soil microbes and rhizosphere in plants. Misidentification may lead to no release of nutrients or a suboptimal release level. Can an aptamer, when immobilized in a nanoscale film, still attach its target? If an aptamer is unable to attach the target, it might misread the signals, leading to too much or too little delivery of nutrients or even no delivery at all. In addition, the presence of impurities in the nutrient payload may force the aptamer to misread the chemical signals and to bind incompletely with the target, resulting in an inaccurate or partial delivery of nutrients.

2.3 Safety aspects and limitations of nanosensors

Given the tremendous benefits of nanosensors, there is substantial public concern over toxicity and the effect on the environment. Nanosensors are facing several issues, including the development of reproducible calibration tools, fouling and drift, use of preconcentration and separation methods for attaining proper concentration of analyte and avoiding saturation, and integration of nanosensors with other elements in a reliable manner. As nanosensors are a relatively new technology, many unresolved nano-toxicology concerns currently hinder their agricultural applications. There is very little information about the long-term adverse effects of nanosensors on plants, soil, and finally on humans [33]. Teow et al. [34] identified health and safety issues associated with applications of nanomaterials. Modern techniques show that nanomaterials with greater reactivity and ability to cross-membrane barriers can result in different toxicodynamic and toxicokinetic properties. Many nanomaterials communicate with proteins and enzymes that trigger oxidative stress and reactive oxygen species production, inducing mitochondrial degradation and triggering apoptosis.

3. High-speed stereovision

Many farmers use broadcast spinner spreaders, also known as centrifugal spreaders, due to their larger working width, low cost, robustness, and high spreading efficiency [35]. In the field, various factors affect the distribution of fertilizers, such as spreader settings and physical properties of fertilizers. This distribution should correspond as closely as possible to the needs of the crop. In fact, the use of an inaccurate amount of fertilizer could reduce the efficiency of production [36]. For example, cereal lodging due to excess nitrogen input significantly reduces profit [37]. To spread the correct amount of fertilizer at the right place in the field, the correct spreader settings are defined by conducting a calibration test taking into account the properties of both the system and the fertilizer. In most cases, the particles of the fertilizer are collected and weighed in standardized trays. As this is a long and challenging method, several alternative techniques have been developed to more efficiently characterize the spreading process and calibrate spreaders [36]. These hybrid techniques first assess the ejection parameters of the fertilizer particles (direction, velocity, and in some cases even size) and second utilize a ballistic flight model to calculate their field landing points.

Among these approaches, those using an imaging system are the most promising because these approaches help calculating all the parameters of ejection without interfering with the flow of fertilizers. Hijazi et al. [38] developed a new method for estimating the movement of fertilizer particles inspired by the technique of particle image velocimetry. This method yielded precise results on the dynamics of particles. Because it is based on a 2D imaging system, the method can only be used for particles that move parallel to the image sensor. However, in real cases, particles make some angle with the sensor plane when spread with a rotating disk spreader. Normally, the particle flight trajectory reaches upto 15 degree in the horizontal plane of the sensor. Even if a flat disk spreader is used, the particles still have a chance of making 4.18 degree angle in the horizontal plane. A 3D method would, therefore, improve the precision of measuring the motion of particles leaving the rotating disk of the spreaders for fertilizer.

Hijazi et al. [35] developed a 3D imaging technique using a high-speed binocular stereovision system in combination with appropriate image processing algorithms to accurately determine the particle parameters leaving centrifugal fertilizer spreader spinning disks. The stereomatching algorithm was validated with a virtual 3D stereovision simulator. The results revealed an error of less than 2 pixels for 90% of fertilizer particles. The system has been tested using an experimental spreader's cylindrical spread pattern. Between the experimental results and the simulated spread pattern, obtained with the developed setup, a 2D correlation coefficient of 90% and a relative error of 27% was found. The developed image acquisition and processing algorithms in combination with a ballistic flight model can make it possible to quickly determine and evaluate the spread pattern that can be used as a tool for spreader design and precise machine calibration.

4. Optical sensors

Scientists and plant producers worldwide use crop canopy sensors to assess crop nutrient status, estimate crop biomass output and yield capacity, breed and select new crops, detect crop stress and disease infestation, suggest fertilizer, and prescribe fertilizer and chemical applications [17]. Crop sensors take advantage of plant optical characteristics and related vigor and health properties. The reason most plants appear green is that they contain chlorophyll, a green pigment that absorbs light in the visible light spectrum's red and blue regions and reflects green light. Less vigorous, unhealthy plants have less chlorophyll and appear less green. A small portion of the infrared region between the visible and microwave regions of the light spectrum is known as near-infrared. As near-infrared has longer wavelengths than visible light, it has unique properties that can be used for remote sensing applications such as detecting nutrient stress, water stress, and weed and pest infestations. High reflection in the near-infrared region is because no plant pigments absorb near-infrared light. The near-infrared light interacts with spongy mesophyll cells on traveling through the plant tissues. Approximately half of its energy is transmitted through the biomass of plants and the other half is reflected. Crop sensors are developed to accurately measure the ratio of absorbed and reflected light levels of specific wavelengths, such as red and near-infrared. The working principle of a crop sensor is illustrated in Fig. 8.3. The data on numerical reflection is exported as vegetative indices. The normalized difference vegetative index (NDVI) is among the most commonly used plant sensing indices. NDVI is closely linked with the amount of plant-produced green biomass or vegetation. NDVI values for bare soil or unhealthy plants range from 0.2 to 0.4 and, for green, vibrant and healthy plants range from 0.5 to 0.9.

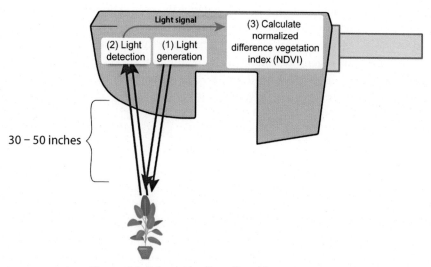

Figure 8.3 Schematic of working of a crop sensor.

The light-emitting diode in Fig. 8.3 generates a mixture of visible red and invisible near-infrared light, which is projected onto the plants. The sensors detect the light reflected from the plants. The integrated electronic circuit transforms the reflectance information into NDVI index values, an index that is connected to the green biomass of a plant [39]. Dunn et al. [40] evaluated nondestructive handheld optical sensors for monitoring the nitrogen status in Maverick Red. A controlled release fertilizer was used as a source of nitrogen in this study. The greenhouse-grown plants were treated with 0, 4, 8, 10, and 12 g of fertilizer. Individual plants are screened for the NDVI and soil-plant analysis creation (SPAD) over eight separate sampling dates beginning 7 days after the application of fertilizer treatment. The width, height, number of umbels, number of flowers, and nitrogen concentration of the leaf were also recorded. For both NDVI and SPAD, linear and quadratic patterns have been observed. In 12 g treatment, plant height and width were the maximum, but in 8 g and 10 g treatments there was no difference. The number of flowers in 10 g treatment was largest, but in 8 and 12 g treatments it was not significant. The number of umbels among fertilizer treatments was not significantly different, but they were all larger than the control.

Franzen et al. [41] revealed that due to poor soil testing relationship to plant response, the prediction of sulfur deficiency is difficult. It was reported that nitrogen sufficient area, established as a standard for an optical sensor, can also be used to detect sulfur deficiency in corn. The red standardized differential vegetation index (NDVI) and the red edge NDVI values were reported using two optical sensors. There were the lowest NDVI readings in the high nitrogen treatment and the highest NDVI readings in the control treatments. A gypsum solution containing 22 kg of sulfur ha^{-1} was applied within 24 h. The sites were revisited 7 days after the application of sulfur. For recording red NDVI and red edge NDVI, the optical sensors were again used. All sites had the highest NDVI readings for the high nitrogen treatment and the lowest NDVI readings for the control treatment. These experiments indicate that high treatment with nitrogen will increase the severity of maize sulfur deficiency. If a lower reading of NDVI is registered in a high application area for nitrogen compared to the surrounding area, there may be a sulfur deficiency.

Fig. 8.4 illustrates a schematic of a setup of the photosensor. This consists of collimated laser modules passing through a macro cuvette where the reaction occurs before a photodiode detects it. Comparing absorption at two wavelengths, the nitrates and nitrites were discriminated. Such measurements can be performed with a high frequency in small batches. Such an embodiment makes this device suitable for in situ real-time determination of nitrates and nitrites, as in the agritech industry's intelligent fertilization applications. By combining the specificity of chemical reactions with the high-resolution power of spectroscopy, this groundbreaking detection approach can outperform existing detection strategies. Knoblauch et al. performed a standardized greenhouse

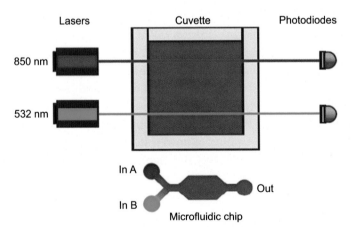

Figure 8.4 Illustration of a macro cuvette-based setup, which can be used to work in a microfluidic chip.

experiment to explore the potential of spectral measurements in a permanent grassland to examine yield response, mineralization of nitrogen, and N_2O emissions [42]. Under controlled environmental conditions, ryegrass was subjected to four levels of mineral fertilizer input over 100 days. Soil temperature and moisture content were tracked automatically, and N_2O and carbon dioxide emission levels were frequently detected. NDVI and simple ratio were somewhat correlated with an increase in biomass as affected by fertilization level. In addition, nonlinear response to elevated fertilizer levels was noticed by increasing N_2O emissions. These results verified the usefulness of optical sensors for the detection of biomass and demonstrated the difficulty in linking spectral measurements of plant traits to nitrogen processes in the soil, despite the latter affecting the former.

5. Remote sensing

There are also remote sensing systems that use aircraft, satellites, and various types of drones to monitor crop conditions, in addition to in-field sensors for monitoring crops [16]. Aerial imagery is being used to assess the volatility of nitrogen status in the US corn belt. Color and infrared aerial photography are used to capture fields for nitrogen deficiency at critical growth stages. The pictures are then analyzed to identify areas that appear to be deficient in nitrogen. The georeferenced images can be analyzed with GIS software along with yield maps, soil maps, and other data to assess potential points for additional scouting, sampling, or application of nitrogen. For aerial photographs, nitrogen stress occurs as a higher reflection in the green and red wavelengths and sometimes in blue. Excessive rainfall can cause fertilizer and soil nitrogen loss, leading to a deficiency of nitrogen. Table 8.1 shows some major plant nutrient deficiencies. To transform the aerial photograph into the yield loss diagram, a quantitative relationship between relative greenness of the plant and yield loss can be used. This information about yield loss could be used to assess the economic impact of nitrogen deficiency, in support of a decision about whether to apply additional nitrogen fertilizer.

Mee et al. [14] discussed various remote sensing techniques specifically related to nutrient stress that could have been affected by other stressors, including an infestation of pests disease infection and water deficit. Such techniques identify and interpret shapes and patterns of remotely sensed imagery based on their respective spectral signatures to measure and

Table 8.1 Representation of some important nutrient deficiencies in plants.

Nutrient	Symptoms	Presence of leaf margin necrosis	Presence of chlorosis	Color and shape of leaf
Nitrogen	On all leaves	No	Yes	Leaf veins, yellowing of leaves
Phosphorous	On older leaves	No	No	Purplish patches
Potassium	On older leaves	Yes	Yes	Yellow patches
Magnesium	On older leaves	No	Yes	Interveinal chlorosis in maize and rice and yellow patches in oil palm
Sulfur	Young leaves	No	Yes	Yellow patches
Manganese, Iron	Young leaves	No	Yes	Interveinal chlorosis
Molybdenum, Zinc, Boron	Young leaves	—	—	Deformed leaves

represent the plant's morphological and physiological changes in response to different stresses. Remote sensing tools tested in this study have generally been shown to be effective and efficient in detecting and tracking nutrient stress from plants, but each has advantages and disadvantages that render them inaccurate when used alone. Nevertheless, in conjunction with other geospatial methods, the prospect of using them is expected to resolve the limitations and increase their ability to achieve more reliable results.

5.1 Imaging and nonimaging chlorophyll fluorescence

Chlorophyll fluorescence can produce fast and accurate data about most plant stresses based on the fluorescence emission patterns of leaves, tissues, and even all plants [43]. The fluorescence emission is detected when parts of light energy, absorbed for photosynthesis by plant chlorophyll, are reemitted with UV-A near 340—360 nm or blue-green light. The latter offers detailed information about its ability to reach deeper tissue layers instead of UV light that is normally intercepted on the epidermal surface. This is because very little UV radiation can pass through the green meso-phyll cells containing chlorophyll pigments, which easily absorb blue and red photons. The corresponding fluorescence ratios are the simultaneous variations between the wavelengths, namely blue, green, and red or more commonly called as the blue-green and red spectra, each of which is defined in Table 8.2. Solar-induced light excitation typically produces blue and green fluorescence between the spectral regions of 440 and 550 nm, while blue-green light excitation provides a red region of 650–800 nm of fluorescence.

Table 8.2 Characteristics of plant-emitted blue-green and red fluorescence [44].

Blue-green fluorescence

Fluorescing pigments Extraction Location and origin Excitation	The ferulic acid mainly develops covalent bonding with cell walls, modulation by cinnamic acids and flavonoids in vacuoles • After alkaline hydrolysis, cell wall-bound ferulic acid • Soluble cinnamic acids and flavonoids with methanol • Signals coming mostly from leaf epidermis (cell walls) • Interaction with soluble phenols (vacuoles) **UV radiation**

Fluorescence characteristics

Emission range Maxima Stress indication	400–570 nm Near 440–450 nm and a shoulder near 520–530 nm • Decrease or increase in the ratio of blue fluorescence (440–690 nm) to red chlorophyll fluorescence (440–740 nm) • In some plants, green fluorescence increases

Chlorophyll fluorescence

Fluorescing pigment Extraction Location and origin Excitation by	Chlorophyll Extracted using organic solvents Leaf mesophyll cells (chloroplasts) Green, red, and blue light or red laser, UV-A radiations, blue laser, UV-laser, green laser, Nd:YAG laser, and pulsed flash lamp

Fluorescence characteristics

Emission range Maxima Short-term stress Long-term stress	650–800 nm 690 nm and 730–740 nm • Changes in fluorescence induction kinetics at the inhabitation of photosynthesis. • Increase in band maxima by 30% • Decrease in chlorophyll content • Increase in maxima more than 30%

Many research works on nonimaging fluorescence of chlorophyll has produced very promising results [43,44]. The nonimaging fluorescence provides only point data measurement with limited information on a sensed small portion of the leaf instead of the entire area of the leaf or canopy, where the more advanced chlorophyll imaging is intended to overcome. Additional fluorescence signatures presented by the multipixel feature of the larger-scale fluorescence sensing with imaging permits thorough screening of all points of leaf. This benefit detects slight changes in fluorescence emission patterns due to a variety of plant internal factors that cannot be detected with nonimaging techniques and therefore eliminates measurement errors. Cadet and Samson [44] used fluorescence ratios in sunflower to discriminate potassium, nitrogen, and phosphorus scarcity. In winter wheat, these fluorescence ratios were also used to differentiate between nitrogen deficiency and disease infection. Fluorescence imaging is also being applied in irrigation management, disease control, and fertilization due to its ability to sense the plant stress caused by water deficiency, nutrient deficiency, and disease infection. In field analysis, fluorescence imaging has some drawbacks due to variations in light intensity coupled with varying site conditions that tend to produce results that differ from laboratory tests. However, experiments with chlorophyll imaging that run the risk of producing incomparable and even contradictory results unless the measuring protocols and parameters used are coordinated. It is also reported that, relative to several fluorescence sensors, measuring distance can have a significant effect on the reflected light.

5.2 Thermography

Different from fluorescence imaging, without an illumination source, thermography can visualize stomatal movement. The thermal signal being studied is a temperature change that is captured in the form of reflected or emitted radiation from the scanned plant. The thermal intensity is measured by the ambient temperature, where the radiation intensity increases with the temperature. The adjustment in the temperature of the leaf usually involves opening and closing stomata for exchanging or cooling energy. The mechanism of cooling by transpiration causes stomatal opening, resulting in lower temperature with ambient heat loss. However, the availability of nutrients in the soil and water flow within the plant is determined by transpiration and ultimately stomatal regulation. Water or nutrient shortages typically interfere with the transportation of dissolved nutrients and water from the soil to root and eventually to the entire plant where nutrient absorption is impaired by higher soil nutrient concentrations [43]. As a result, the stomata close to avoid further loss of moisture and raise the temperature of the surface of the plant. Chaerle et al. [45] revealed that magnesium starved bean plant showed higher leaf temperature on thermal imaging under controlled conditions. Tilling et al. [45] recorded relatively higher temperature than well-fertilized barley with nitrogen as the reference nutrient for fertilized barley.

Although thermography is typically passive in nature, it can be active when a stimulus such as incident light is introduced to induce a targeted subject's temperature change. Active thermography enables monitoring of the internal heterogeneity of the leaf in relation to disease-induced change or development, whereas passive thermography evaluates similar capacity changes by estimating water evaporation. Thermography can detect the stress of the plant, but it does not have the ability to distinguish between the stressors. It should, therefore, be used in combination with other sensing techniques, such as chlorophyll imaging, to simultaneously monitor and differentiate different stresses [45].

5.3 Multispectral and hyperspectral imaging

Multispectral devices measure reflection in 40 nm wide bands in red, green, blue, and near-infrared areas and can be expanded to a maximum of 10 wavelengths [46]. This seems to be the greatest distinction between strategies of hyperspectral and multispectral techniques. The hyperspectral method uses additional spectral bands that result in higher spectral resolution or a narrower wavelength of approximately 10 nm or less in the visible and near-infrared band spectrum, offering detailed information covering the larger functional and structural properties of the plant. Compared to thermography, in combination with chlorophyll fluorescence, the hyperspectral technique can be used to track and discern the various plant internal responses to stress based on changes in photosynthetic efficiency and spectral properties. It has been shown that plant stress at various levels across multiple crops is effectively sensed using hyperspectral classification such as water content, disease, and crop nutrient status. Because of the wider wavelengths available in hyperspectral imaging, it is important to filter the unnecessary bands and focus on the distinct ones with the highest optical sensitivity to a particular plant stress to avoid redundancy in the generated data and images. The filtering process can eliminate noise from the environment to produce more accurate spectral data. For example, in characterizing the distribution of nitrogen, phosphorus, and potassium in rapeseed leaves, Zhang et al. [47] stated that spectral reflection at certain wavelengths is more effective. Although reflectance changes due to nutrient stress were successfully identified using the hyperspectral method, their study suggested further data processing to overcome the uncertainty arising from overlapping spectral bandwidths.

5.4 Nutritional status monitoring with satellites and airborne platforms

Continuous monitoring of crop conditions is important for tracking over time dynamics of crop growth and development. The research offers timely information that can help identify problem areas impacted by various vegetative factors including water status, distribution of nutrients and possible disease, as well as weed encroachment that can only occur over longer periods of time. In doing so, field operations such as fertilizer application and pesticide recommendations can be adjusted in terms of timing and application rate to accommodate the different growing requirements of crops at different times over the growing period for increased agricultural productivity and food supply [48]. The demand for plant nutrients is usually diverse across various stages of growth. It has clearly shown that plant nitrogen status constantly changes over the entire period of growth and that the fertilization strategies should act in response to these changes. Plant reliance on the availability of nitrogen in natural soil is impractical because it is subject to soil type, previous crop management, and environment at that time. Therefore, long-term monitoring records are required for farmers to track crop yield trends and assess their resilience against changing climatic conditions, which periodically alter the distribution of rainfall and temperature fluctuations.

Remote monitoring of plant conditions and growth prediction can be accomplished by integrating their multiple image data with correct process-based simulation models using satellite and aircraft platforms. Data from moderate resolution imaging spectroradiometer have been used to forecast crop yields on selected crops, including barley, field peas, rapeseed, and spring wheat. Herrmann and his team measured vegetation changes in wheat and potato by estimating the leaf area index with new micro-spacecraft and sentinel-2 [49]. Although satellite sensors are useful in most plant condition assessments, due to their lower spatial and spectral resolutions and longer revisit period, they have limited use in precision farm management. In recent years, technological advances have resulted in airborne sensors with higher spatial and spectral resolutions and shorter revisit times. However, some glitches remain unsolved, such as substantial data processing time and higher acquisition costs, making them less effective on a large scale, both operationally and economically.

6. Conclusions

Field scouting and visual plant assessments are routine work of most of the crop producers. This practice provides valuable information on crop growth, development, and nutrient status as well as the pressures of pests, weeds, and diseases. Such knowledge is used to recommend management practices such as herbicide treatments and water controls for irrigation. The severe stresses caused by the nutrient deficiencies, plant disease, water shortage, or pest infestations can easily be identified through the naked eye. However, plant stress issues that are more subtle or just developing may not be as obvious and may go unnoticed or misinterpreted. Several techniques have been developed to efficiently characterize the nutrients' spreading process and monitoring of the crop growth parameters. Among these approaches, those using an imaging system are the most promising because it is possible to monitor the growth parameters and concentration of nutrients without interfering with the crop. The crop canopy sensors are assumed best to assess crop nutrient status, crop biomass yield, output capacity, breed, crop stress, and disease infestation. There are also remote sensing systems that use aircraft, satellites, and various types of drones to monitor crop conditions, in addition to in-field sensors for monitoring crops. Fluorescence imaging is also being applied in irrigation management, disease control and fertilization due to its ability to sense the plant stress caused by water deficiency, nutrient deficiency, and disease infection. In field analysis, fluorescence imaging has some drawbacks due to variations in light intensity coupled with varying site conditions that tend to produce results that differ from laboratory tests. Different from fluorescence imaging, without an illumination source, thermography can visualize stomatal movement. Although thermography is typically passive in nature, it can be active when a stimulus such as incident light is introduced to induce a targeted subject's temperature change. Active thermography enables monitoring of the internal heterogeneity of the leaf in relation to disease-induced change or development, whereas passive thermography evaluates similar capacity changes by estimating water evaporation.

Nanomaterials contribute to the diagnostic selectivity and usability by having a larger surface area for aptamer immobilization and by offering the system its own optophysical and electrochemical properties. Nevertheless, the creation of field-applicable nanosensor technologies still poses several obstacles. Further research into the possible solutions to these problems and the creation of innovative nanomaterials can be hoped to improve the construction of an inexpensive and conveniently operable nanomaterial dependent sensing device. Therefore, the full potential of nanosensors is yet to be understood in the agricultural sector. Future research should be focused on developing innovative and effective nanoscale products, processes, and smart devices to realize their full potential in agriculture and the effects on humans and the environment.

References

[1] J.N. Galloway, A.R. Townsend, J.W. Erisman, M. Bekunda, Z. Cai, J.R. Freney, et al., Transformation of the nitrogen cycle: recent trends, questions, and potential solutions, Science 320 (2008) 889—893.

[2] P. Glibert, J. Harrison, C. Heil, S. Seitzinger, Escalating worldwide use of urea-a global change contributing to coastal eutrophication, Biogeochemistry 77 (2006) 441—463.

[3] M. Naz, S. Sulaiman, Attributes of Natural and Synthetic Materials Pertaining to Slow-Release Urea Coating Industry, vol. 33, 2016.

[4] K.A. Ibrahim, M.Y. Naz, S. Shukrullah, S.A. Sulaiman, A. Ghaffar, N. AbdEl-Salam, Controlling nitrogen pollution via encapsulation of urea fertilizer in cross-linked corn starch, BioResources 14 (4) (2019) 7775.

[5] E.W. Boyer, C.L. Goodale, N.A. Jaworski, R.W. Howarth, Anthropogenic nitrogen sources and relationships to riverine nitrogen export in the Northeastern USA, Biogeochemistry 57/58 (2002) 137—169.

[6] M.E. Trenkel, Improving Fertilizer Use Efficiency. Controlled-Release and Stabilized Fertilizers in Agriculture, The International Fertilizer Industry Association, Paris, 1997.

[7] C. Tolescu, H. Iovu, Polymer conditioned fertilizers, UPB Sci. Bullet., Ser. B 72 (2) (2010).

[8] A. Shaviv, Controlled Release Fertilizers, in: IFA International Workshop on Enhanced-Efficiency Fertilizers, Frankfurt, International Fertilizer Industry Association Paris, France, 2005.

[9] M.E. Engelsjord, O. Fostad, B.R. Singh, Effects of temperature on nutrient release from slow-release fertilizers, Nutr. Cycl. Agroecosyst. 46 (3) (1996) 179—187, https://doi.org/10.1007/bf00420552.

[10] S. Jin, Y. Wang, J. He, Y. Yang, X. Yu, G. Yue, Preparation and properties of a degradable interpenetrating polymer networks based on starch with water retention, amelioration of soil, and slow release of nitrogen and phosphorus fertilizer, J. Appl. Polym. Sci. 128 (1) (2012) 407—415.

[11] D.S. Saleh, M. Hemati, Experimental study and modeling of fluidized bed coating and agglomeration, Powder Technol. 130 (2003) 116—123.

[12] S. Butzen, Common nitrogen fertilizers and stabilizers for corn production, Crop Insights 23 (2) (2013).

[13] G.R. Hergert, C. Ferguson, C. Wortmann, Enhanced Efficiency Fertilizers: Will They Enhance My Fertilizer Efficiency? Proceedings of the 3rd Annual Crop Production Clinics, University of Nebraska-Lincoln Entension, 2011.

[14] C.Y. Mee, S.K. Balasundram, A.H.M. Hanif, Detecting and monitoring plant nutrient stress using remote sensing approaches: a review, Asian J. Plant Sci. 16 (2017) 1—8.

[15] G.N. Rameshaiah, J. Pallavi, S. Shabnam, Nano fertilizers and nano sensors — an attempt for developing smart agriculture, Int. J. Eng. Res. Gen. Sci. 3 (1) (2015) 314—320.

[16] H.F. Reetz, Fertilizers and Their Efficient Use, International Fertilizer Industry Association, Paris, France, 2016.

[17] O. Walsh, Nitrogen Management in Field Crops with Reference Strips and Crop Sensors, University of Idaho, Moscow, Russia, 2015. BUL 896.

[18] L.C. Qin, Electron diffraction from carbon nanotubes, Rep. Prog. Phys. 69 (2006) 2761—2821.

[19] M.H. Jnr, J.L. Vicente, Kinetic study of liquid-phase Adsorptive removal of heavy metal ions by almond tree (Terminalia catappa L.) leaves waste, Bull. Chem. Soc. Ethiop. 21 (3) (2007) 349—362.

[20] A. Aqel, K.M.M.A. El-Nour, R.A.A. Ammar, A. Al-Warthan, Carbon nanotubes, science and technology part (I) structure, synthesis and characterisation, Arab. J. Chem. 5 (1) (2012) 1—23, https://doi.org/10.1016/j.arabjc.2010.08.022.

[21] A.H. Barber, R. Andrews, L.S. Schadler, H.D. Wagner, On the tensile strength distribution of multiwalled carbon nanotubes, Appl. Phys. Lett. 87 (20) (2005) 203106, https://doi.org/10.1063/1.2130713.

[22] Y. Cheng, J. Zhang, Y.Z. Lee, B. Gao, S. Dike, W. Lin, et al., Dynamic radiography using a carbon-nanotube-based field-emission x-ray source, Rev. Sci. Instrum. 75 (10) (2004) 3264—3267, https://doi.org/10.1063/1.1791313.

[23] J. Liu, Y. Zhang, Z. Zhang, Study on application of nanometer biotechnology on the yield and quality of winter wheat, J. Anhui. Agric. Sci. 35 (2008) 15578—15580.

[24] Y.-f. Qian, C.-h. Shao, C.-f. Qiu, X.-m. Chen, S.-l. Li, W.-d. Zuo, C.-r. PENG, Primarily study of the effects of nanometer carbon fertilizer synergist on the late rice, Acta Agric. Boreali-Sinica S2 (2010).

[25] N.M. Yatim, A. Shaaban, M.F. Dimin, F. Yusof, J.A. Razak, Effect of functionalised and non-functionalised carbon nanotubes-urea fertilizer on the growth of paddy, Trop. Life Sci. Res. 29 (1) (2018) 17—35, https://doi.org/10.21315/tlsr2018.29.1.2.

[26] J.-J. Adjizian, R. Leghrib, A.A. Koos, I. Suarez-Martinez, A. Crossley, P. Wagner, et al., Boron- and nitrogen-doped multi-wall carbon nanotubes for gas detection, Carbon 66 (2014) 662—673, https://doi.org/10.1016/j.carbon.2013.09.064.

[27] J. Kim, S.-W. Choi, J.-H. Lee, Y. Chung, Y.T. Byun, Gas sensing properties of defect-induced single-walled carbon nanotubes, Sensor. Actuator. B Chem. 228 (2016) 688—692, https://doi.org/10.1016/j.snb.2016.01.094.

[28] T.-G. Cha, B.A. Baker, M.D. Sauffer, J. Salgado, D. Jaroch, J.L. Rickus, et al., Optical nanosensor architecture for cell-signaling molecules using DNA aptamer-coated carbon nanotubes, ACS Nano 5 (5) (2011) 4236—4244, https://doi.org/10.1021/nn201323h.

[29] Z. Majeed, N.K. Ramli, N. Mansor, Z. Man, A comprehensive review on biodegradable polymers and their blends used in controlledrelease fertilizer processes, Rev. Chem. Eng. 31 (1) (2015) 69—95.

[30] M.Y. Naz, S.A. Sulaiman, Slow release coating remedy for nitrogen loss from conventional urea: a review, J. Contr. Release 225 (2016) 109—120, https://doi.org/10.1016/j.jconrel.2016.01.037.

[31] Y. Sultan, R. Walsh, C. Monreal, M.C. DeRosa, Preparation of functional aptamer films using layer-by-layer self-assembly, Biomacromolecules 10 (5) (2009) 1149—1154, https://doi.org/10.1021/bm8014126.

[32] M.C. DeRosa, C. Monreal, M. Schnitzer, R. Walsh, Y. Sultan, Nanotechnology in fertilizers, Nat. Nanotechnol. 5 (2) (2010), https://doi.org/10.1038/nnano.2010.2, 91-91.

[33] A. Dubey, D.R. Mailapalli, Nanofertilisers, nanopesticides, nanosensors of pest and nanotoxicity, in: E.L. Agriculture (Ed.), Sustainable Agriculture Reviews 19, Sustainable Agriculture Reviews, Springer International Publishing Switzerland, 2016.

[34] Y. Teow, P.V. Asharani, M.P. Hande, S. Valiyaveettil, Health impact and safety of engineered nanomaterials, Chem. Commun. 47 (25) (2011) 7025−7038, https://doi.org/10.1039/C0CC05271J.

[35] B. Hijazi, S. Cool, J. Vangeyte, K.C. Mertens, F. Cointault, M. Paindavoine, J.G. Pieters, High speed stereovision setup for position and motion estimation of fertilizer particles leaving a centrifugal spreader, Sensors 14 (11) (2014) 21466−21482.

[36] F.J. García-Ramos, A. Boné, A. Serreta, M. Vidal, Application of a 3-D laser scanner for characterising centrifugal fertiliser spreaders, Biosyst. Eng. 113 (1) (2012) 33−41.

[37] H. Van Grinsven, H. Ten Berge, T. Dalgaard, B. Fraters, P. Durand, A. Hart, et al., Management, regulation and environmental impacts of nitrogen fertilization in northwestern Europe under the Nitrates Directive: a benchmark study, Biogeosciences 9 (12) (2012) 5143−5160.

[38] B. Hijazi, J. Vangeyte, F. Cointault, J. Dubois, S. Coudert, M. Paindavoine, J. Pieters, Two-step cross correlation-based algorithm for motion estimation applied to fertilizer granules' motion during centrifugal spreading, Opt. Eng. 50 (6) (2011) 067002.

[39] B.S. Tubaña, D.B. Arnall, O. Walsh, B. Chung, J.B. Solie, K. Girma, W.R. Raun, Adjusting midseason nitrogen rate using a sensor-based optimization algorithm to increase use efficiency in corn (Zea mays L.), J. Plant Nutr. 31 (8) (2008) 1393−1419. https://doi.org/10.1080/01904160802208261.

[40] B.L. Dunn, A. Shrestha, C. Goad, Determining nitrogen fertility status using optical sensors in geranium with controlled release fertilizer, J. Appl. Hortic. 17 (1) (2015) 7−11.

[41] D.W. Franzen, L.K. Sharma, H. Bu, A. Denton, Evidence for the ability of active-optical sensors to detect sulfur deficiency in corn, Agron. J. 108 (2016) 2158−2162, https://doi.org/10.2134/agronj2016.05.0287.

[42] C. Knoblauch, C. Watson, C. Berendonk, R. Becker, N. Wrage-Mönnig, F. Wichern, Relationship between remote sensing data, plant biomass and soil nitrogen dynamics in intensively managed grasslands under controlled conditions, Sensors 17 (7) (2017) 1483.

[43] A. Ač, Z. Malenovský, J. Olejníčková, A. Gallé, U. Rascher, G. Mohammed, Meta-analysis assessing potential of steady-state chlorophyll fluorescence for remote sensing detection of plant water, temperature and nitrogen stress, Rem. Sens. Environ. 168 (2015) 420−436, https://doi.org/10.1016/j.rse.2015.07.022.

[44] É. Cadet, G. Samson, Detection and discrimination OF nutrient deficiencies IN sunflower BY blue-green and chlorophyll-a fluorescence imaging, J. Plant Nutr. 34 (14) (2011) 2114−2126, https://doi.org/10.1080/01904167.2011.618572.

[45] L. Chaerle, I. Leinonen, H.G. Jones, D. Van Der Straeten, Monitoring and screening plant populations with combined thermal and chlorophyll fluorescence imaging, J. Exp. Bot. 58 (4) (2006) 773−784, https://doi.org/10.1093/jxb/erl257.

[46] D.J. Mulla, Twenty five years of remote sensing in precision agriculture: key advances and remaining knowledge gaps, Biosyst. Eng. 114 (4) (2013) 358−371, https://doi.org/10.1016/j.biosystemseng.2012.08.009.

[47] J. Zhang, R. Pu, W. Huang, L. Yuan, J. Luo, J. Wang, Using in-situ hyperspectral data for detecting and discriminating yellow rust disease from nutrient stresses, Field Crop. Res. 134 (2012) 165−174, https://doi.org/10.1016/j.fcr.2012.05.011.

[48] W. Wang, X. Yao, Y.-c. Tian, X.-j. Liu, J. Ni, W.-x. Cao, Y. Zhu, Common spectral bands and optimum vegetation indices for monitoring leaf nitrogen accumulation in rice and wheat, J. Integr. Agric. 11 (12) (2012) 2001–2012, https://doi.org/10.1016/S2095-3119(12)60457-2.

[49] I. Herrmann, A. Pimstein, A. Karnieli, Y. Cohen, V. Alchanatis, D.J. Bonfil, LAI assessment of wheat and potato crops by VENμS and Sentinel-2 bands, Rem. Sens. Environ. 115 (8) (2011) 2141–2151, https://doi.org/10.1016/j.rse.2011.04.018.

CHAPTER 9

Trends and technologies behind controlled-release fertilizers

Reshma Soman, Subin Balachandran
School of Biosciences, Mahatma Gandhi University, Kottayam, Kerala, India

In the previous chapters, we have understood about the controlled-release fertilizers. This chapter will go into detail about the trends and technologies behind CRFs.

To lower the environmental, ecological, and health hazards and improve the nutrient supply of crops, CRFs play a major role as they are fertilizer granules embedded within a matrix of carrier molecules which can control the release of nutrients into the soil. Generally, CRFs when applied to the soil, on attaining suitable conditions, get activated and slowly get released into the soil. This helps in the controlled release of fertilizer without harming the ecosystem.

1. Introduction

Conventional fertilizers cause a lot of environmental problems like eutrophication leading to oxygen depletion, thereby causing the death of fishes; along with that, it also causes nasty odor. Hence to rectify these environmental problems, controlled-release fertilizers (CRFs) are trending in our ecosystem. CRFs help to improve the nutrient supply to crops as they are fertilizer granules incorporated within carrier molecules in which different parameters such as rate of release, pattern, and time duration of release can be predicted. It has great importance in agriculture as they can supply nutrients to crops through a single application for an entire season [34]. Due to the high costs of fertilizers and low cost for crops, traditional CRFs are not economically feasible. Therefore new economical CRFs are important in agriculture [3].

CRFs can be prepared from different components including gums or resins, biomaterials, seed mucilage, clay, and organoclays. According to the compositions, the mode of action of CRFs varies and also its relevance. CRFs are ecofriendly and have been in use for many years. Owing to their ability for gel formation and release of fertilizer to the environment, different formulations are still on research for CRFs. Nanomaterials are

Controlled Release Fertilizers for Sustainable Agriculture
ISBN 978-0-12-819555-0
https://doi.org/10.1016/B978-0-12-819555-0.00009-1

widely used in research and fertilizer incorporated with nanoparticles as nanofertilizers are trending in day-to-day life. Nanofertilizers are encapsulated with different nutrients including nitrogen, phosphorus, zinc, silica, and so on and are added to the crops as per their requirements.

Another trend of CRF formulation is by the use of superhydrophobic nanoparticles. Those nanoparticles having the property of water repellency are under research for the preparation of CRFs. These nanoparticles become magnetic nanoparticles on induction with a magnet and then finally form a matrix. This property made them useful as a CRFs.

In the future, superhydrophobic magnetic nanoparticles can be prepared from more cost-effective pig fat and be of great use. Nanofertilizers had already governed the agriculture and new technologies for the preparation of nanofertilizers can be adopted in the future.

2. Generalized mechanism of controlled-release fertilizers

Plants require both macromolecules and micromolecules for their growth. Nitrogen (N) is an essential and adequate element for the growth of a plant and nitrogen fertilizers are of great demand in the horticulture production. Due to the improved N use efficiency and hiking N prices, the need for better N fertilizers is increasing in our day-to-day life.

The N release mechanism from a slow-release fertilizer of urea-formaldehyde (UF) was demonstrated by Guo and his colleagues in 2006. When the fertilizer is released into the field, upon the favorable condition, water gets absorbed from the soil thereby making the coating materials to become swollen and finally transforming them into hydrogels. The hydrogels further lead to the alteration of the 3D network of coating materials by increasing the pore size of the coating material. This helps in diffusing the fertilizer to the deep into the gel network and also the formation of a water layer between the core molecule and coat material that was swollen due to water absorption. The water-soluble part of the UF core molecule is dissolved, as the water slowly gets penetrated into the cross-linked polymer network. By the interchange of water in the soil and water in the hydrogel formed, through the swollen network, the soluble part of the fertilizer slowly oozes out and gets released into the soil. This helps the microorganisms in the soil to penetrate the fertilizer through the swollen coatings, accumulate together around the UF core granule, and disintegrate the insoluble part of N into urea and ammonia thereby slowly releasing it into the soil with the help of dynamic exchange.

3. Compositions of controlled-release fertilizers

Controlled-release fertilizers can be of different types according to their carrier molules [34], Table. 9.1.

3.1 Natural polymers

Synthetic polymers are used mainly for the preparation of fertilizers, and these are replaced by natural polymers as they are environmentally friendly, having the capability of controlling soil erosion, and are inexpensive [34]. Natural polymers, on regarding their composition can be divided into three: (i) gum, (ii) seed polysaccharide, and (iii) bio-derived materials.

3.1.1 Gum

Hydrophobic or hydrophilic gums can be extracted from different plants. These gums or resins possess gel formation capacity and on combining with water forms hydrogels thereby making their use in CRF formulations [34].

Table 9.1 Different compositions of controlled release fertilizers.

3. Compositions of controlled-release fertilizers

3.1. Natural polymers
3.1.1 Gum
3.1.1.1 Xanthan gum
3.1.1.2 Gaur gum
3.1.1.3 Gellan gum
3.1.1.4 Mucuna gum
3.1.1.5 Gum gopal
3.1.1.6 Karaya gum
3.1.2 Seed polysaccharide
3.1.2.1 Tamarind seed polysaacharide
3.1.2.2 *Mimosa pudica* seed mucilage
3.1.2.3 *Leucaena leucocephala* seed polysaccharide
3.1.3 Bio-derived materials
3.1.3.1 Chitosan
3.1.3.2 Carrageenan
3.1.3.3 Pectin
3.2. Modified clays
3.3. Other components

3.1.1.1 Xanthan gum

Hydrophilic water-soluble high molecular weight xanthan gum is an anionic bacterial heteropolysaccharide that is also incorporated into a carrier molecule to produce CRFs [14,21,34].

3.1.1.2 Gaur gum

Apart from the cationic and anionic hydrophobic polysaccharide gaur gum is a nonionic water-soluble polysaccharide having detergent properties, which is isolated from *Cyamopsis tetragonolobus* seeds. It can be used as CRF but the use is yet to be exploited [9,14,20,34].

3.1.1.3 Gellan gum

The fermentation product obtained from *Pseudomonas elodea* is an anionic water-soluble polysaccharide having high molecular weight heteropolymers consisting each of β-D-glucuronic acid, α-L-rhamnose, and two repeating units of β-D—glucose all together forming a tetrasaccharide repeating unit polysaccharide. It also forms hydrogels by penetrating the core molecule thereby making it one of the important formulations of CRFs [15,34].

3.1.1.4 Mucuna gum

The cotyledons of *Mucuna hagillepesa* plant can be used for the extraction of mucuna gum. It is composed of D-mannose and D-glucose with D-galactose making it a biodegradable amorphous polymer. It exhibits rapid drug release as it is devoid of cross-linking and can be used as CRFs [11,34].

3.1.1.5 Gum copal (GC)

Gum Copal is a water–insoluble (hydrophilic) naturally occurring resin isolated from *Bursera bipinnata*. It exhibits diffusion of water, swelling, and degradation of fertilizer granules (excipients) thereby making it a zero–order kinetic model [34].

3.1.1.6 Karaya gum

It is a hydrophilic polymer that has the property of forming a hydrogel. It is obtained naturally from *Sterica usens* composed of three sugar monomers—glucuronic acid, rhamnose, and galactose—and can be widely used in CRF production [20,34].

3.1.2 Seed polysaccharide
Different seeds having the capability to form matrix when embedded with fertilizers can be used for the preparation of CRFs with seed polysaccharide compositions. Mainly they can be of different types:

3.1.2.1 Tamarind seed polysaccharide (TSP)
TSP is composed of three monomers of sugar—galactose, xylose, and glucose—in the molar ratio 1:2:3. The galactoxyloglucan isolated from tamarind species *T. indica* have nontoxic, biocompatible, and economical agro-based material which can be used as CRF formulations considering their properties [14,20,34].

3.1.2.2 *Mimosa pudica* seed mucilage
Hitherto, the CRF industry is investigating the use of *Mimosa pudica* seed in the production of CRF formulation as studies proved that bioadhesion time of *Mimosa* mucilage polymer can obtain more than 85% release of drug in 10 h. *Mimosa* mucilage has a matrix-forming ability that can be exploited for making formulations [14,20,34].

3.1.2.3 *Leucaena leucocephala* seed polysaccharide (LLSP)
Similar to that of *Mimosa pudica*, LLSP is also a galactoxyloglucan which can form hydrophilic gum. Intercalating *L. leucocephala* residues with nitrogen fertilizer can be used for the combinational release of hydrophobic and hydrophilic drugs in a controlled way. The use of *L. leucocephala* residues improves the time interval of nitrogen released into the soil [8,14,34,41].

3.1.3 Bio-derived materials
Heteropolymers which can be obtained from shells, seaweeds, and fruits can be used for the preparation of gels. These heteropolymers react with water to form hydrogels and this property aids in the preparation of CRFs [34].

3.1.3.1 Chitosan
Chitosan is a heteropolymer mainly obtained by deacetylation of chitin of crustacean shells partially. Glucosamine and N-acetyl glucosamine are the copolymers of chitosan; at the same time, chitin of invertebrates, insects, and yeasts are rich in chitosan. Chitosan can be used for the preparation of hydrogels and nanoparticles. These chitosan nanoparticles can be incorporated into the NPK fertilizers to use it as CRFs [7,20,34].

3.1.3.2 Carrageenam

High molecular weight natural polymer can be extracted from red seaweed which is comprised of galactose and 3,6-anhydrogalactose monomers. Since it has the property for the formation of gel, it can be used for the preparation of hydrogels and thereby used as CRFs [10,12—14,34].

3.1.3.3 Pectin

Pectin is a component of fruits like apple, grapes, gooseberry, and also can form pectin-based hydrogel and thus used as CRFs [14,20,24,34].

3.2 Modified clay

Nanoparticles made of clay having large surface area and reactivity of nanolayers can be used in the CRF preparations. Nanoclay apart from the nanosize structures can act as an active substrate for physicochemical and biological reactions. These properties make it an important carrier in the manufacture of CRF formulations [2,34].

3.3 Organoclay chemistry

Organoclays are clay minerals with less than 2 μm particle size, possessing crystal lattice unit. Their high specific surface area, hydrogel capacity, charge, and crystallike structure, along with colloidal particle size, are used for CRF formulation production [4,23,25,34].

4. Failure of release

CRFs are of great importance; apart from that, they sometimes cause failures on administration to the soil. Raban (1994) explained in his study about the reasons for release failures of fertilizers. The commencement of the release process leads to the slow penetration of water through the coating. Water vapor condenses and leads to the dissolving of fertilizer. It causes internal pressure inside the coated granule, leading to rupture of the coating by increasing the internal pressure over the threshold value and finally destroys the coating, and there is continuous release of the fertilizer into the soil. Thus, the thought of new technologies in the controlled release of fertilizers came to the limelight [34].

5. Petroleum-based polymers

Superhydrophobic characteristic is the water-repelling character shown by many species found in nature to avoid the penetration of water droplets into them. Water-repelling characteristic of lotus leaves are well known as it can lower surface energy by ascribing the micro—nanoscale heaves intercalated within the microstructures. Slow-release fertilizer coated with superhydrophobic petroleum-based polymer is now trending in agriculture and can be used to improve the agricultural production.

Several natural polymers mentioned above are used against petroleum-based coating materials. But the mechanical behavior and poor hydrophobic characters of the mentioned coating materials affect the nutrient requirement of crops by lowering the longevity of nutrient release. Water-insoluble hydrophobic substances are used on the mentioned polymer-coated materials which can ameliorate the water repellency of the coated materials helping as a superhydrophobic treatment.

The superhydrophobic surfaces of SBSF should be constructed for the preparation of superhydrophobic slow-release fertilizer. Using the external magnetic field produced, superhydrophobic surfaces are constructed by the self-assembly owing to the help of superhydrophobic modified magnetic sensitive nanoparticles Fe_3O_4 (SMN). The magnetic interaction between SMN and walls of coating machine (made up of ferrous iron) induce SMN to get uniformly dispersed on the outermost coating surface, thereby constructing a superhydrophobic microstructure of SBSF. This can be made in future with the help of long-chain fatty acid as large-scale production while pig fat can be used to produce ecofriendly and green biopolymer coating of SBSF without aiding any byproducts.

The SBSF thus formed have long-lasting (durable) properties. Its slow release longevity is more than 100 days making it exhibit a better slow-release performance juxtaposing to that of other biopolymers which are unmodified [17].

6. Nano fertilizers

Nanotechnology is an emerging technology mainly in the field of medicine thereby yielding nanomedicine; similarly, revolutionary change using nanoparticles is a breakthrough in agricultural development. Nanoparticles can be used to intercalate with fertilizer as a carrier molecule for the preparation of nanofertilizers; they have very small molecule size thereby having a large surface area-to-volume size ratio. These properties of nanoparticles aid for the exceptional properties of nano fertilizers including optoelectronic and physicochemical properties. Nanofertilizers are used for the efficient uptake of fertilizers by plants by slowly releasing the fertilizer ingredient. They help to enhance the biological health of the soil, reduce the demand for fertilizers giving high productivity, and save natural resources. Nanofertilizers, as they are in nanoform, will be efficient in the agriculture as the fertilizer can be easily taken up by the plant while bulk fertilizer contains large particles of ionized salts that make nutrient absorption by the plant difficult [26,27,36].

7. Nanofertilizers are of different types

(a) Nitrogen (N)
(b) Phosphorous (P)
(c) Zinc (Zn)
(d) Iron oxide (Fe_3O_4)
(e) Silica (Si)
(f) Titanium (Ti)
(g) Carbon-based nanoparticles

7.1 Nitrogen (N) nanofertilizers

Conventional nitrogen fertilizers are widely used in agriculture. Nitrogen that is applied into the field can be lost by different ways including leaching of NO_3 and surface runoff to water bodies, through the emission of N in the form of N_2O or NO and vaporization of N as ammonia [5]. The escape of nitrogen in these ways causes serious environmental hazards. To wrap up this serious problem, nitrogen nanofertilizer can be used in the field.

Nitrogen nanofertilizers are nanoscale fertilizers of N having a size of less than 200 nm and a concentration of 50 kg per hectare. This was a specialized formulation developed by Sri Lankan Institute of Nanotechnology by using urea coated hydroxyapatite as carrier molecules for the preparation of N nano fertilizer which has the capability to slow–release N and targeted delivery to the soil. As urea and hydroxyapatite are rich in N and P, usage of this these for the preparation of nanoparticles will make the soil rich in both minerals. These fertilizers can be exposed to the soil as a mode of treatment mainly to rice plants [6,19]. The N nanofertilizer thus applied will aid to the N slow release into the soil, thereby curing the problem of environmental hazards and also aiding in slow uptake of N by the plant resulting in improved rice yield [19]. The weight of nanohybrid synthesized was 40% lesser the weight of N and on comparing to urea, the release of N is 12 times slower making it provide a better yield at 50% lower concentration of urea.

7.2 Phosphorus (P) nano fertilizers

Phosphorus is very essential for plants. Rock phosphate was used as a keen ingredient in the preparation of conventional fertilizer. Global rock phosphate is estimated to have only 100 more years to be lasting in the world as a result of this; a hike in the P fertilizer price is a serious issue in the global market. Apart from the hike in cost of fertilizer, P fertilizer on application to the soil undergoes chemical bond with other elements present in the soil in the form of calcium—phosphorus (Ca—P), magnesium—phosphorus (Mg—P), aluminum—phosphorus (Al—P), iron—phosphorus (Fe—P), and zinc — phosphorus (Zn—P). These chemical bonds of P make the plant's uptake of P very difficult due to the unavailability of free P and also the excess P gets fixed into the soil. These are harmful to the environment as well as to nature; to resolve the issue, P nanofertilizer can be used as a remedy [32]. Richardson in 2001 [40] reported that just as nitrogen-fixing bacteria, phosphate solubilizing bacteria, different types of fungi that produce phytase enzyme and phosphatases, and many other microorganisms including those which can produce organic acids have thoroughly undergone investigation for improving the P uptake by plants. Unfortunately, the study did not go well, limiting the phosphate solubilization. Tarafdar et al. in 2012 [48] tried to make fungal nanoparticles for the P uptake by using tricalcium phosphate as its precursor salt and the nanoparticles possess about 28 nm in size and contain 62% of P. The idea of nanoparticle preparation came in the ideology of Liu and Lal and they researched on the same. Liu and Lal in 2017 reported that they were able to synthesize cellulose stabilized nanoparticles having a size of 16 nm and they successfully did their research in the soybean plant. They also reported that the treated soybean show a hike in phonological growth by 33% thereby increasing the yield to 18%. This is because the hydroxyapatite nanoparticles form less chemical bonds with the elements in soil and possess a weaker interaction with them; this helps to increase the P uptake and also yield of a plant.

Even though there are several studies reported in the formation of fungal and bacterial nanoparticles, the way to increase P uptake by plants is still not known. To reveal the mystery and to make the plants take up more P, now Zinc nanoparticles are being used for the preparation of P nanofertilizer as carrier molecules. Zn nanoparticles contain Zn which is one of the important cofactors of phosphate and phytase. The supplementation of Zn nanoparticles-based P nanofertilizer thus in one hand helps for the easy uptake of P by plant and in another way provides Zn (which is also an essential element) to plant. These studies were done by Raliya et al. [28,30,31,38] and in their lab review [32] they described that the Zn nanoparticles, when given as foliar application to the plant, had helped to increase the enzyme activity (phytase and phosphatase) and resulted in about 11% uptake of P by leguminous plants and cereals. This increases the growth of plants, biomass, the nutritive value of cereals and legumes, as well as the yield of the crop. Venkatachalam et al. in 2017 [52] reported that Zn nanoparticles P nanofertilizers possess the ability to increase antioxidant properties of plants; they also increased the photosynthetic pigments and biomass of plants.

8. Synthesis of nanoparticle or nanofertilizer

Nanoparticles are particles of nanosize hence the synthesis of nanoparticles are very important. Zahra et al. [18,28,31,37,44], and Raliya et al. [34] made nanoparticles by wet methods. The wet method adopted were sol-gel, homogenous preparation, enzyme biosynthesis method, hydrothermal methods, protein template methods, and reverse micelle methods. The dry methods for the preparation of nanoparticles were adopted by Li et al. [24](Jiang et al. [18], Storbel and Pratsinis [43], An et al. [1], and [40, 43]. The dry methods they used were aerosol-based processes by using single element nanoparticles, oxide semiconductors, metals, metal oxides, metal alloys, polymers, and doped andand composite nanoparticles. The nanoparticles thus prepared should have low cost and capability of producing in bulk. There are several advantages and disadvantages according to the synthesis method of a nanoparticle. The gas-phase aerosol preparation of nanoparticle can be prepared in a single step, have better control over the particle size and shape. They have passivation of the surface; they possess monodispersity for a few percents and are scalable. This can be used for the synthesis of controlled nanocomposite but as a disadvantage, the large aggregate may be formed by the preparation.

The liquid phase nanoparticle synthesis method can be both biological and chemical. The chemical liquid phase nanoparticle synthesis has precise control on morphology of metal nanoparticles but it has comparatively lesser biocompatibility due to the surface coating with harmful chemicals. The environmental friendly biological methods have a surface coating with natural micro-macromolecules and are highly biocompatible methods. But as they require natural resources, the rate of molecule synthesis is low with wide polydispersity and poly shape.

The solid-phase physical method is rapid and is scale-up synthesis. But this method has a broad polydispersity and poly shape [32].

9. Conclusion

Conventional fertilizers are harmful to the environment and health, and are also costly. To rectify these problems, CRFs are being used in the soil. Even the CRFs cause some problems for the environment. The ideology of superhydrophobic CRF and usage of nanotechnology for the production of nanofertilizer according to the plant need thus came to limelight and is now gaining attention as a trending technology. Superhydrophobic CRF has water-repelling capacity and these fertilizers are made with the help of a magnet. While nanofertilizers are CRFs having a particle size in nanoform intercalated within a carrier molecule which has a good effect on the soil. The nanofertilizer can be of different types according to plant need. The mode of delivery of nano fertilizers depends on the plant and the carrier molecule, as well as the mineral incorporated within it to make the fertilizer. Nanofertilizers on applying to the field slowly get released and finally increase the yield and productivity of the crop and soil, respectively, without harming nature. In short, new trends and technologies are adopted to lower the harmful effects of the fertilizers and increase the productivity of the soil by the slow release capacity of the CRFs

References

[1] W.J. An, E. Thimsen, P. Biswas, Aerosol-chemical vapor deposition method for synthesis of nanostructured metal oxide thin films with controlled morphology, J. Phys. Chem. Lett. 1 (2010) 249−253.

[2] B.B. Basak, S. Pal, S.C. Datta, Use of modified clays for retention and supply of water and nutrients, Curr. Sci. (2012) 1272−1278.

[3] A.D. Blaylock, J. Kaufmann, R.D. Dowbenko, Nitrogen fertilizer technologies, in: Western Nutrient Management Conference, vol. 6, March 2005, pp. 8−13.

[4] C. Aguzzi, P. Cerezo, C. Viseras, C. Caramella, Use of clays as drug delivery systems: possibilities and limitations, Appl. Clay Sci. 36 (1−3) (2007) 22−36.

[5] S.R. Carpenter, N.F. Caraco, D.L. Correll, R.W. Howarth, A.N. Sharpley, V.H. Smith, Nonpoint pollution of surface waters with phosphorus and nitrogen, Ecol. Appl. 8 (1998) 559−568.

[6] M. Chhowalla, Slow-release nano fertilizers for bumper crops, ACS Cent. Sci. 3 (2017) 156.

[7] E. Corradini, M.R. De Moura, L.H.C. Mattoso, A preliminary study of the incorporation of NPK fertilizer into chitosan nanoparticles, Express Polym. Lett. 4 (8) (2010).

[8] S.K. Das, G.S. Reddy, K.L. Sharma, K.P.R. Vittal, B. Venkateswarlu, M.N. Reddy, Y.V.R. Reddy, Prediction of nitrogen availability in soil after crop residue incorporation, Fert. Res. 34 (3) (1993) 209—215.

[9] V.N. Deshmukh, S.P. Singh, D.M. Sakarkar, Formulation and evaluation of sustained-release metoprolol succinate tablet using hydrophilic gums as release modifiers, Int. J. Pharm. Tech. Res. 1 (2) (2009) 159—163.

[10] K. Eswaran, P.K. Ghosh, A.K. Siddhanta, J.S. Patolia, C. Periyasamy, A.S. Mehta, et al., U.S. Patent No. 6,893,479, U.S. Patent and Trademark Office, Washington, DC, 2005.

[11] P. Gill, T.T. Moghadam, B. Ranjbar, Differential scanning calorimetry techniques: applications in biology and nanoscience, J. Biomol. Tech. 21 (4) (2010) 167.

[12] H. Hezaveh, I.I. Muhamad, Controlled drug release via minimization of burst release in pH-response kappa-carrageenan/polyvinyl alcohol hydrogels, Chem. Eng. Res. Des. 91 (3) (2013) 508—519.

[13] H. Hosseinzadeh, Controlled release of diclofenac sodium from pH-responsive carrageenan-g-poly (acrylic acid) superabsorbent hydrogel, J. Chem. Sci. 122 (4) (2010) 651—659.

[14] S.E. Hanafi, M. Ahmad, Physical and chemical characteristics of controlled release compound fertiliser, Eur. Polym. J. 36 (10) (2000) 2081—2088.

[15] T. Jamnongkan, S. Kaewpirom, Controlled-release fertilizer based on chitosan hydrogel: phosphorus release kinetics, Sci. J. Ubu. 1 (1) (2010) 43—50.

[16] J. Jiang, D.R. Chen, P. Biswas, Synthesis of nanoparticles in a flame aerosol reactor with independent and strict control of their size, crystal phase and morphology, Nanotechnology 18 (2007) 285603.

[17] J. Xie, Y. Yang, B. Gao, Y. Wan, Y.C. Li, D. Cheng, T. Xiao, K. Li, Y. Fu, J. Xu, Q. Zhao, Y. Zhang, Y. Tang, Y. Yao, Z. Wang, L. Liu, Magnetic-sensitive nanoparticle self-assembled superhydrophobic biopolymer-coated slow-release fertilizer: fabrication, enhanced performance and mechanism, ACS Nano (2019). February, 28.

[18] R. Kaul, P. Kumar, U. Burman, P. Joshi, A. Agrawal, R. Raliya, J. Tarafdar, Magnesium and iron nanoparticles production using microorganisms and various salts, Mater. Sci. Poland 30 (2012) 254—258.

[19] N. Kottegoda, C. Sandaruwan, G. Priyadarshana, A. Siriwardhana, U. Rathnayake, D. Berugoda Arachchige, A. Kumarasinghe, D. Dahanayake, V. Karunaratne, G. Amaratunga, Urea-hydroxyapatite nanohybrids for slow release of nitrogen, ACS Nano 11 (2017) 1214—1221.

[20] S. Kumar, S.K. Gupta, Natural polymers, gums and mucilages as excipients in drug delivery, Polim. Med. 42 (3—4) (2012) 191—197.

[21] L. Wu, M. Liu, Preparation and properties of chitosan-coated NPK compound fertilizer with controlled-release and water-retention, Carbohydr. Polym. 72 (2) (2008) 240—247.

[22] S. Li, Y. Ren, P. Biswas, S.D. Tse, Flame aerosol synthesis of nanostructured materials and functional devices: processing, modelling, and diagnostics, Prog. Energy Combust. Sci. 55 (2016) 1—59.

[23] M.E. Parolo, L.G. Fernandez, I. Zajonkovsky, M.P. Sánchez, M. Baschini, Antibacterial activity of materials synthesized from clay minerals, in: Science against microbial pathogens: Communicating Current Research and Technological Advances, Microbiology Series, vol. 3, Formatex, 2011, pp. 144—151.

[24] M.R. Guilherme, A.V. Reis, A.T. Paulino, T.A. Moia, L.H.C. Mattoso, E.B. Tambourgi, Pectin-based polymer hydrogel as a carrier for release of agricultural nutrients and removal of heavy metals from wastewater, J. Appl. Polym. Sci. 117 (6) (2010) 3146—3154.

[25] R.E. Grim, Clay Mineralogy, International Series in the Earth and Planetary Sciences, McGraw-Hill, New York, NY, USA, 1968.

[26] R. Raliya, V. Saharan, C. Dimkpa, P. Biswas, Nano fertilizer for precision and sustainable agriculture: current state and future perspectives, J. Agric. Food Chem. 66 (26) (2017) 6487—6503.

[27] R. Raliya, P. Biswas, Environmentally benign bio-inspired synthesis of Au nanoparticles, their self-assembly and agglomeration, RSC Adv. 5 (2015) 42081—42087.

[28] R. Raliya, I. Rathore, J. Tarafdar, Development of microbial nano factory for zinc, magnesium, and titanium nanoparticles production using soil fungi, J. Bionanoscience 7 (2013) 590—596.

[29] R. Raliya, J. Tarafdar, Biosynthesis and characterization of zinc, magnesium and titanium nanoparticles: an eco-friendly approach, Int. Nano Lett. 4 (2014) 1—10.

[30] R. Raliya, J.C. Tarafdar, P. Biswas, Enhancing the mobilization of native phosphorus in the mung bean rhizosphere using ZnO nanoparticles synthesized by soil fungi, J. Agric. Food Chem. 64 (2016) 3111—3118.

[31] R. Raliya, J.C. Tarafdar, K. Gulecha, K. Choudhary, R. Ram, P. Mal, R. Saran, Review article; scope of nanoscience and nanotechnology in agriculture, J. Appl. Biol. Biotechnol. 1 (2013) 041—044.

[32] R. Raliya, V. Saharan, C. Dimkpa, P. Biswas, Nano fertilizer for precision and sustainable agriculture: current state and future perspectives, J. Agric. Food Chem. 66 (2018) 6487—6503.

[33] A.E. Richardson, Prospects for using soil microorganisms to improve the acquisition of phosphorus by plants, Funct. Plant Biol. 28 (2001) 897—906.

[34] S.I. Sempeho, H.T. Kim, E. Mubofu, A. Hilonga, Meticulous overview on the controlled release fertilizers, Adv. Chem. 2014 (2014).

[35] R. Strobel, S.E. Pratsinis, Flame aerosol synthesis of smart nanostructured materials, J. Mater. Chem. 17 (2007) 4743—4756.

[36] L.V. Subbaiah, T.N.V.K.V. Prasad, T.G. Krishna, P. Sudhakar, B.R. Reddy, T. Pradeep, Novel effects of nanoparticulate delivery of zinc on growth, productivity, and zinc biofortification in maize (zeamays l.), J. Agric. Food Chem. 64 (2016) 3778—3788.

[37] J.C. Tarafdar, S. Sharma, R. Raliya, Nanotechnology: Interdisciplinary science of applications, Afr. J. Biotechnol. 12 (2013) 219—226.

[38] J. Tarafdar, R. Raliya, H. Mahawar, I. Rathore, Development of zinc nanofertilizer to enhance crop production in pearl millet (Pennisetum americanum), Agric. Res. 3 (2014) 257—262.

[39] J. Tarafdar, R. Raliya, I. Rathore, Microbial synthesis of phosphorous nanoparticle from tri-calcium phosphate using *Aspergillus tubingensis* tfr-5, J. Bionanoscience 6 (2012) 84—89.

[40] T. Tsuzuki, P.G. McCormick, Mechanochemical synthesis of nanoparticles, J. Mater. Sci. 39 (2004) 5143—5146.

[41] V. Pendyala, C. Baburao, K.B. Chandrasekhar, Studies on some physicochemical properties of Leucaena Leucocephala bark gum, J. Adv. Pharm. Technol. Research 1 (2) (2010) 253—259.

[42] P. Venkatachalam, N. Priyanka, K. Manikandan, I. Ganeshbabu, P. Indiraarulselvi, N. Geetha, K. Muralikrishna, R.C. Bhattacharya, M. Tiwari, N. Sharma, S.V. Sahi, Enhanced plant growth-promoting role of phycomolecules coated zinc oxide nanoparticles with p supplementation in cotton (*Gossypium hirsutum* L.), Plant Physiol. Biochem. 110 (2017) 118—127.

[43] W.-N. Wang, Y. Jiang, P. Biswas, Evaporation-induced crumpling of graphene oxide nanosheets in aerosolized droplets: confinement force relationship, J. Phys. Chem. Lett. 3 (2012) 3228—3233.

[44] Z. Zahra, M. Arshad, R. Rafique, A. Mahmood, A. Habib, I.A. Qazi, S.A. Khan, Metallic nanoparticle (TiO$_2$ and Fe$_3$O$_4$) application modifies rhizosphere phosphorus availability and uptake by Lactuca sativa, J. Agric. Food Chem. 63 (2015) 6876—6882.

Further reading

J. Cao, Y. Feng, X. Lin, J. Wang, Arbuscular mycorrhizal fungi alleviate the negative effects of iron oxide nanoparticles on bacterial community in rhizospheric soils Front, Environ. Sci. 4 (2016) 10.

C.O. Dimkpa, P.S. Bindraban, J. Fugice, S. Agyin-Birikorang, U. Singh, D. Hellums, Composite micronutrient nanoparticles and salts decrease drought stress in soybean Agron, Sustain. Dev. 37 (2017) 5.

R. Liu, R. Lal, Synthetic apatite nanoparticles as a phosphorus fertilizer for soybean (Glycine max), Sci. Rep. 4 (2015) 5686.

M. Guo, M. Liu, R. Liang, A. Niu, Granular urea-formaldehyde slow-release fertilizer with superabsorbent and moisture preservation, J. Appl. Polym. Sci. 99 (6) (n.d.), pp.3230.

R. Raliya, J. Tarafdar, Biosynthesis of gold nanoparticles using Rhizoctonia bataticola TFR-6, Adv. Sci. Eng. Med. 5 (2013) 1073—1076.

R. Raliya, J.C. Tarafdar, ZnO nanoparticle biosynthesis and its effect on phosphorous-mobilizing enzyme secretion and gum contents in clusterbean, (Cyamopsis tetragonoloba l.)Agric. Res. 2 (2013) 48—57.

S. Raban, E. Zeidel, A. Shaviv, J.J. Mortwedt, A. Shaviv, Release mechanisms controlled-release fertilizers in practical use, in: Proceedings of the 3rd International Dahlia Greidinger Symposium on Fertilization and the Environment, 1997, pp. 287—295.

J.C. Tarafdar, R. Raliya, Rapid, low cost and ecofriendly approach for iron nanoparticle synthesis using Aspergillus oryzae tfr9J, Nanopart 2013 (2013) 14.

B.L. Vallee, K.H. Falchuk, The biochemical basis of zinc physiologyPhysiol, Rev 73 (1993) 79—118M.

P. Wang, E. Lombi, F.-J. Zhao, P.M. Kopittke, Nanotechnology: a new opportunity in plant sciences, Trends Plant Sci 21 (2016) 699—712.

CHAPTER 10

Nanotechnology in controlled-release fertilizers*

Rakhimol K.R.[1], Sabu Thomas[1], Nandakumar Kalarikkal[1], Jayachandran K.[2]

[1]International and Inter University Centre for Nanoscience and Nanotechnology, Mahatma Gandhi University, Kottayam, Kerala, India; [2]School of Biosciences, Mahatma Gandhi University, Kottayam, Kerala, India

1. Introduction

The major challenge in the fertilizer research is to improve the efficiency and availability of fertilizers to plants by enhancing the availability of existing one or by developing a new fertilizer with high efficiency. Discovery of controlled-release fertilizers (CRFs) improved the fertilizer efficiency by the sustained release of fertilizer. CRFs are mainly prepared through the physical or chemical modification of conventional fertilizers by reducing their solubility and mobility [1].

Common ways of fertilizer application is the spraying and foliar application. It leads to the loss of fertilizers without reaching to the soil and plant. So, repeated application is needed. This led to the discovery and development of controlled and slow release fertilizers with high efficiency and accuracy.

CRFs can be produced with unique characteristics. Mainly it is achieved through (a) the coating or encapsulation of conventional fertilizer with a porous material or (b) by reducing or controlling the water penetration into the fertilizer. Nanotechnology is a new field of research which includes the synthesis of materials in nano regime (1−100 nm) with unique physical and chemical properties and their application in different fields including agriculture. Agriculture is the occupation of about 60% of the people

---

* This chapter reports on types of nano-based fertilizers, methods of preparation, advantages of applying them, and nanotoxicity. From this point of view, it can be useful to scientists interested in enhancing fertilizer availability to plant, preventing fertilizer wastage, as well as preventing leaching of nutrients into soil and water bodies.

Although, studies on application of nanotechnology in agriculture are scanty, a number of important reports in the review were not referenced. Figures and tables were not mentioned in the text. The other comments are in the paper.

Controlled Release Fertilizers for Sustainable Agriculture
ISBN 978-0-12-819555-0
https://doi.org/10.1016/B978-0-12-819555-0.00010-8

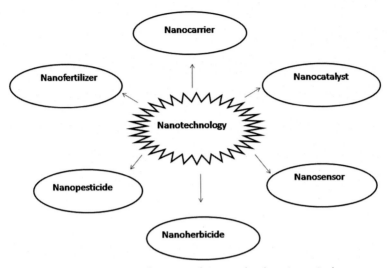

Figure 10.1 Major applications of nanotechnology in agriculture.

worldwide. Nanotechnology has the potential to modernize the agriculture industry through the development of nanofertilizers, sensors to detect and manage diseases, improved plant growth, etc.

Nanobiotechnology is the merging of nanotechnology with biotechnology to produce better agriculture and food products for improved living. "Nanocatalyst" is a novel term to increase fertilizer efficiency. It will also reduce the dose of fertilizers. It prevents the wastage of fertilizer in soil by helping the plants to take up maximum amount (Fig. 10.1).

The major challenges in the agriculture sector are optimized production of agriculture products from a limited area without any adverse effect on environment and sustainability. Applications of nanotechnology to obtain the maximum result with minimum side effects include

(a) Enhancing the growth and productivity of the plant using nanofertilizers.
(b) Development of nanopesticides and nanoherbicides for the plant protection.
(c) Controlled release of agrochemicals using nanocarriers.
(d) Periodic detection of the development using nanosensors.

2. Designing of nanofertilizer

Nowadays nanotechnology has more practical applications in the field than experimental or theoretical approach. Nanotechnology offers nanostructures which can act as carriers or vectors for controlled release. Nanotechnology could build "smart fertilizers" to protect the environment and enhance the fertilizer efficiency. Nanofertilizers combined with nanosensors play a major role in synchronizing the nutrient release from the fertilizer and nutrient uptake by the plant. By this method, the fertilizer concentration for application can be optimized, as well as leaching of nutrients into soil and water bodies can be prevented [2].

Encapsulation of fertilizer in nanocarrier can be performed in three different ways. They are

(a) Encapsulation of nutrient in materials with nanopores,
(b) Coating of nutrients using thin polymer films,
(c) Converting the nutrients to nanoscale dimensions.

3. Advantages of nanofertilizers

3.1 Controlling the release profile of nutrients

Nanofertilizers or nutrients encapsulated in nanoporous materials could intelligently control the release of entrapped nutrients from its core. It could optimize the release according to the plant's need and environmental conditions.

3.2 Enhancing the solubility and bioavailability

Nanoformulations could enhance the solubility of the nutrient ingredients in water and thus enhance the availability of those nutrients to the plants. The direct uptake of nutrients by plants could result in the reduction of soil absorption and fixation.

3.3 Enhanced release duration

Controlled and slow release of nutrients from the nanocarriers without any wastage could enhance the duration of nutrient release. Most of them can provide nutrients to the plants in its entire lifetime by changing the release profile according to the plants' development stages.

4. Types of nanofertilizers

4.1 Layered double hydroxides in controlled release

LDH is a two-dimensional material also known as anionic clays. It has many layers in its structure and in the interlayer spacing of LDH, anionic materials can be intercalated. It favors the release of anions in a controlled manner [3,4]. There is no restriction for any anions to balance the positive

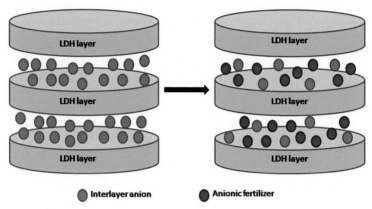

Figure 10.2 Anionic exchange in layered double hydroxide.

charge of LDH. Depending upon the composition and synthetic methods, it has diverse properties. Some of them are capability for anion exchange and thermal stability.

The most important and risky task in fertilizer release research is the maintenance of nitrogen and phosphorous in soil in the form of nitrate and phosphate. Nitrates will be easily denitrified and the phosphate-based fertilizers lose their efficiency due to the phosphate fixation by soil [5]. Hence, it is essential to develop a method for the sustainable and efficient release of nitrogen and phosphorous. LDH is a better option for this. LDH allows the intercalation of nitrogen and phosphorous in its interlayer space by ion exchange between LDH and nitrates or phosphates [6] (Fig. 10.2).

LDH and their mixed oxide byproducts can be employed as adsorbents to take up the anionic species which contaminate the environment, soil, and water. They are commonly used to remove the unused and waste nitrates and phosphates and other pesticides from soil and water through adsorption and released back to the plants in a controlled manner by recycling [7,8].

4.2 Zeolites

Zeolites are aluminosilicate minerals with microporous structure. The porous structures of the zeolite can accommodate cations in it by adsorption and ion exchange [9,10]. "Clinoptilolite" (type of zeolite which is naturally occurring) is commonly used for the controlled release of potassium as per the plant's need. It can also be used to release nitrogen by loading ammonia in it. Zeolites can absorb water up to half of their weight and release slowly

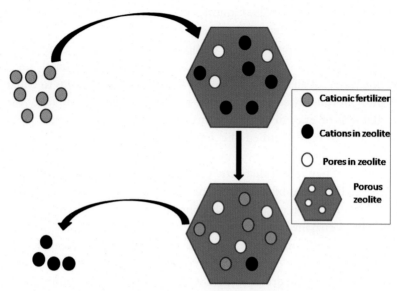

Figure 10.3 Cationic exchange in zeolites.

depending on the water content of the soil. It can prevent the root rot of the plant. Tsintskaladze et al. have developed a nanonitrogenous zeolite to supply nitrogen to the plant in its entire life cycle. They found that it reduces the wastage of nitrogen and water to the soil by providing them as per the plant's need. Flores et al. synthesized potassic zeolites and studied their agricultural application. It can be used as potassium fertilizer to plants [11] (Fig. 10.3).

Studies on the comparison of nitrogen leakage from the urea fertilizers and the zeolite coated urea fertilizers suggest that about 65% reduction in the nitrogen leakage was observed when the urea was coated with zeolite [12]. Zeolite nanocomposites are a novel material which expresses the controlled release efficiency greater than that of zeolite nanoparticles. They have a higher water retention and absorption capacity than zeolite nanoparticles. It was found to enhance the seed germination and plant growth. It can deliver the fertilizer in the different development stages of the plant as is required. It can also enhance flowering and fruiting of plants [13].

4.3 Nanoclays

Phosphorous fixation by soil is one of the most important limitations in agriculture sector. It gradually reduces the availability of phosphorous to the plant. Hence, for improved productivity, the phosphorous fertilizers need to be introduced repeatedly. Continuous use of the chemicals has adverse effect on the environment and is costly [14].

Mainly, clay has been used to immobilize the enzymes to improve the catalytic activity of them [15]. Nanoclay is a good material for the immobilization of enzymes which catalyze organic phosphorous to enhance the phosphorous availability to the plants and help to uptake them. The stability of enzymes also enhances after the immobilization [16,17]. Calabi-Floody et al. studied the immobilization of acid phosphatase in nanoclay and their application in controlled release of phosphorous from cattle dung. They found that activity of enzyme and rate of reaction were enhanced. The immobilization of enzymes was mainly through encapsulation. Mineralization of organic phosphorous was also found to be faster in immobilized enzyme treatments [18].

Zinc deficiency in cereals is a widespread problem. Cereals from the fields with low zinc concentration contain low Zn content and result in malnutrition in children [19]. $ZnSO_4 \cdot 7H_2O$ is the commonly used Zn source. But the efficiency is very low because it leaches at very faster rates in soil [20]. A novel method is needed to enhance the efficiency and enrichment of zinc in grains. Mandal et al. examined the nanoclay–polymer bionanocomposites as a controlled release vehicle for zinc. The composites were found to have enhanced zinc release efficiency.

4.4 Carbon-based nanofertilizer

Carbon is an important element, which helps in the promotion of life. Carbon exists in all the organic compounds and enters into biochemical pathways. Instead of this, nanostructured carbon also plays vital roles in plants [21]. Biochar is the leftover biomass of the plants in fields. It contains large number of carbon nanostructures within it. These carbon structures undergo aerial oxidation and form pores in its surface. These pores help to collect micronutrients from the soil for their controlled release in future. The presence of numerous pores in the surface helps to absorb a large amount of water for its controlled release during drought. This will help the soil to stay wet and with essential nutrients for the seed germination and plant growth [22,23].

Ashfaq et al. examined the release of Cu (an essential micronutrient in the plant growth) using carbon nanofiber (CNF) carriers. CNF could hold the CuNPs in its tip. Cu-CNFs improved the germination rate and shoot and root regeneration. It could also enhance the chlorophyll and protein content of each plant. The encapsulation of this Cu-CNF in polymeric membrane will also enhance the release of nutrients [24].

4.5 Mesoporous silica

Mesoporous silica is a new development in the field of nanotechnology. The common applications of mesoporous silica are in the medical field such as bioimaging and drug delivery. Large number of pores present in it allows the entering of desired particle in it. Mesoporous silica nanoparticles are nontoxic and biocompatible with large surface area due to the presence of pores. The pores of silica are stable and tunable [25].

There are many advantages in using mesoporous silica nanoparticles in controlled release fertilizers. They are

(1) Silica is a naturally occurring substance. So, it will not cause any hazardous effect to environment.
(2) Apart from the controlled release of fertilizers, silica is able to protect the fertilizers from their rapid immature degradation.
(3) Absorbed nanosilica by the root forms a thin film around the cell walls and helps to defend against stress.
(4) The ordered and stable pores of mesoporous silica as well as the highly functionalizable surface could control the loading and controlled release of fertilizers.

There are several methods to load the fertilizers into the mesoporous silica nanoparticles. Most common methods are immersion and fluid bed loading. Mesoporous silica nanoparticles are a good candidate to deliver urea fertilizer. Urea is highly photo liable. They will easily hydrolyze in soil. To overcome these issues, an adequate method is needed to protect the urea particle from immature degradation. Silica nanoparticle is able to protect and deliver the fertilizer by tuning the pore size and surface functionalization [26].

4.6 Polymeric nanoparticles

Chitosan is a deacetylated form of chitin and it is extensively used as drug delivery system in medical field. Chitosan is a biocompatible material because the chitin is a naturally occurring material in crustacean shells. It can also act as antibacterial and antifungal agent [27].

The chitosan is a cationic polymer, so it can interact with negatively charged materials. It has additional advantage of tunable size. In agriculture, incorporation of NPK fertilizers can be done for their controlled release. The incorporation of fertilizers in chitosan nanoparticles is also very easy by dissolving them in nanoparticle solution and magnetic stirring [28].

Polydopamine is an organic polymer which can form a thin layered coating with a thickness of nanometer range on any material [29]. Polydopamine is an analog of melanin. It possesses no cytotoxic effect and excellent biocompatibility. The thickness of polydopamine layers can be adjusted by optimizing four controlling factors [30,31]. They are

(a) concentration of the monomer, dopamine,

(b) pH of the solution,

(c) time taken for the deposition,

(d) number of cycles allowed for deposition.

Time taken for biodegradability of the polydopamine can be tuned by altering the dopamine content in each layer of the coating [32]. It is also helpful to control the release of contents entrapped in it [33]. This helps polydopamine to act as a controlled release vehicle for fertilizers [34].

5. Nanosensors

Sensors are sophisticated devices used to respond to biological, electrical, or chemical signals. It converts physical parameter to electrically measured signal [35]. The factors needed to be considered for designing a sensor are

- Environmental conditions of the sensor (temperature or humidity),
- Measurement range of the sensor,
- Calibration of the sensor,
- Repeatability of the instrument under same environmental conditions,
- Accuracy of the sensor.

Nanosensors are the miniatures of the sensors which can do wonders in the field of sensing. Selectivity and sensitivity of the nanosensors are very high than the traditional sensors. Nanosensors are able to sense and report the information simultaneously. The main advantages of nanosensors are

- Highly selective and sensitive,
- Real-time detection,
- Portable,
- Low cost [36].
  Main applications of nanosensors in agriculture are
- Nanosensors could monitor soil pH and moisture.
- Nanosensors attached to nanocarriers could detect the amount of fertilizer release at a specific period of time.
- Nanocapsules for the smart delivery of fertilizers and pesticides.
- Nanosensors for precision farming to help the utilization of natural resources in a more sophisticated way.
- Nanosensors for the warning of upcoming climate changes.
- Nanosensors for the detection and quantification of heavy metal content in the soil.

- Nanochips for the tracking of nutrients in the plants.
- Nanosensors for the detection of soil pathogens which attack the plants.
- Nanosensors to detect the spoilage of fertilizers in the soil.
- Smart sensors for the detection of quality of the agriculture products [37].

6. Methods of fertilizer release

Fertilizer release from nanocarriers is a very important factor which affects the mode of delivery of the fertilizer in the agriculture field. Mainly the fertilizers are released from nanocarriers in three ways.

Fertilizers which are incorporated in the hydroxyapatite or layered double hydroxide systems will be delivered by the ion exchange method. Anionic or cationic fertilizers can be incorporated into these systems according to the ions present in their interlayer spacing. This will be more specific in the fields [6].

Polymer coated fertilizers are released by the degradation of polymer layers in the soil. Biodegradable polymers are more suitable for the same. Release time will be dependent on the layers of the polymer around it [29].

If the fertilizers are incorporated into any nanoporous systems, the release will be happened by the leaching of the fertilizers from the nanoporous system through the pores. The leaching depends on the size and numbers of pores present in the system [25].

7. Factors affecting the designing of nano-based controlled-release system

The designing of controlled delivery systems using nanoparticles is affected by some factors [38]. They are

(1) Physicochemical characteristics of nanoparticles—the main factors which affect the nanofertilizers are the physical and chemical properties of the nanoparticle used. The chemical nature and elemental compositions are the major chemical factors and size, shape, surface roughness are the physical factors which affect the release.

(2) Type of fertilizer—the fertilizer which is to be delivered will affect the designing of its delivery system. Mainly, the charge of the fertilizer is the most affecting factor. Most of the delivery system works on the ion exchange mechanism, so the charge of the encapsulating fertilizer will play a major role in the designing of its carrier.

(3) Type of interaction of fertilizer with the carrier—the type of interaction between fertilizer and carrier is a very important parameter to be tested. If the interaction is through chemical bonding, the delivery will be restricted. The interaction will be physical (entrapment) or through weak chemical bonding (ion exchange) for better delivery.

(4) Mode of delivery—the mode of delivery is to be confirmed before designing a nanofertilizer. Because the delivery will be affected by pH, moisture content, and other environmental factors, the release of fertilizers from their carriers will be affected by their mode of delivery.

(5) Optimization of the release studies—release of the fertilizer from the carrier at the same pH, moisture content, and other conditions of the field soil should be studied in the laboratory conditions to optimize the fertilizer release in the fields and to study the dose of fertilizer to be encapsulated in the carrier.

8. Nanotoxicity in the fields

In spite of the numerous benefits of nanotechnology in agriculture field, the toxicity of the nanoparticle to crops and humans is a major concern among people. The studies on the toxicity of nanomaterials in agriculture and human are limited. The nanoparticles used in agriculture fields can easily enter into the soil system and if it binds with any toxic material, the nanoparticle could carry them into the soil, crops, and into humans [39]. It will cause major health issues in humans. The major problem with the use of nanomaterials is that they can enter into the cells and stay for a long period. Method of elimination of the nanoparticle that has entered into the body is unknown. Because of its small size, the health issues that develop because of their presence may remain undetected for a long time. Migration of nanoparticles from fertilizers to plants and from plants to humans will lead to deposition of them in humans and cause serious health issues. Nanoparticles could cross the blood–brain barrier. They are also very risky to detect and treat.

One of the effective solutions against nanotoxicity is the use of green nanotechnology. By the use of safe and efficient particles, the toxicity to the plants, environment, water bodies, animals, and humans can be minimized [40,41].

9. Conclusion

Nanotechnology plays a key role in many fields including medicine and agriculture. Application of nanotechnology in agriculture is still in its infant stage. Nanotoxicity is a little explored area of research and needs to be further explored. Nowadays, the research is focusing on green nano-technological methods for sustainable agriculture without any harmful side effects. In future modern technologies like nanotechnology-based agriculture products will help to increase the crop yield to the maximum.

References

[1] M.E. Trenkel, Controlled-release and Stabilized Fertilizers in Agriculture, vol. 11, International Fertilizer Industry Association, Paris, 1997.

[2] H. Guo, J.C. White, Z. Wang, B. Xing, Nano-enabled fertilizers to control the release and use efficiency of nutrients, Curr. Opin. Environ. Sci. Health 6 (2018) 77−83.

[3] F. Cavani, F. Trifiro, A. Vaccari, Hydrotalcite-type anionic clays: preparation, properties and applications, Catal. Today 11 (2) (1991) 173−301.

[4] C. Forano, T. Hibino, F. Leroux, C. Taviot-Guého, 1 layered double hydroxides, Dev. Clay Sci. 1 (2006) 1021−1095.

[5] K.J. McInnes, R.B. Ferguson, D.E. Kissel, E.T. Kanemasu, Field measurements of ammonia loss from surface applications of urea solution to bare soil 1, Agron. J. 78 (1) (1986) 192−196.

[6] L.O. Torres-Dorante, J. Lammel, H. Kuhlmann, T. Witzke, H.W. Olfs, Capacity, selectivity, and reversibility for nitrate exchange of a layered double-hydroxide (LDH) mineral in simulated soil solutions and in soil, J. Plant Nutr. Soil Sci. 171 (5) (2008) 777−784.

[7] J. Inacio, C. Taviot-Gueho, C. Forano, J.P. Besse, Adsorption of MCPA pesticide by MgAl-layered double hydroxides, Appl. Clay Sci. 18 (5−6) (2001) 255−264.

[8] L.P.F. Benício, R.A. Silva, J.A. Lopes, D. Eulálio, R.M.M.D. Santos, L.A.D. Aquino, et al., Layered double hydroxides: nanomaterials for applications in agriculture, Rev. Bras. Ciência do Solo 39 (1) (2015) 1−13.

[9] V.S. Marakatti, A.B. Halgeri, G.V. Shanbhag, Metal ion-exchanged zeolites as solid acid catalysts for the green synthesis of nopol from Prins reaction, Catal. Sci. Technol. 4 (11) (2014) 4065−4074.

[10] V.S. Marakatti, A.B. Halgeri, Metal ion-exchanged zeolites as highly active solid acid catalysts for the green synthesis of glycerol carbonate from glycerol, RSC Adv. 5 (19) (2015) 14286−14293.

[11] H. Zhao, G.F. Vance, G.K. Ganjegunte, M.A. Urynowicz, Use of zeolites for treating natural gas co-produced waters in Wyoming, USA, Desalination 228 (1−3) (2008) 263−276.

[12] A. Dubey, D.R. Mailapalli, Zeolite coated urea fertilizer using different binders: fabrication, material properties and nitrogen release studies, Environ. Technol. Innov. 16 (2019) 100452.

[13] A. Lateef, R. Nazir, N. Jamil, S. Alam, R. Shah, M.N. Khan, M. Saleem, Synthesis and characterization of zeolite based nano−composite: an environment friendly slow release fertilizer, Microporous Mesoporous Mater. 232 (2016) 174−183.

[14] F. Matus, X. Amigo, S.M. Kristiansen, Aluminium stabilization controls organic carbon levels in Chilean volcanic soils, Geoderma 132 (1−2) (2006) 158−168.

[15] M.A. Rao, L. Gianfreda, Properties of acid phosphatase−tannic acid complexes formed in the presence of Fe and Mn, Soil Biol. Biochem. 32 (13) (2000) 1921−1926.

[16] D. Moelans, P. Cool, J. Baeyens, E.F. Vansant, Using mesoporous silica materials to immobilise biocatalysis-enzymes, Catal. Commun. 6 (4) (2005) 307−311.

[17] M.C. Floody, B.K.G. Theng, P. Reyes, M.L. Mora, Natural nanoclays: applications and future trends—a Chilean perspective, Clay Miner. 44 (2) (2009) 161—176.

[18] M. Calabi-Floody, G. Velásquez, L. Gianfreda, S. Saggar, N. Bolan, C. Rumpel, M.L. Mora, Improving bioavailability of phosphorous from cattle dung by using phosphatase immobilized on natural clay and nanoclay, Chemosphere 89 (6) (2012) 648—655.

[19] M.A. Oliver, P.J. Gregory, Soil, food security and human health: a review, Eur. J. Soil Sci. 66 (2) (2015) 257—276.

[20] C.M. Monreal, M. DeRosa, S.C. Mallubhotla, P.S. Bindraban, C. Dimkpa, Nano-technologies for increasing the crop use efficiency of fertilizer-micronutrients, Biol. Fertil. Soils 52 (3) (2016) 423—437.

[21] D. Jariwala, V.K. Sangwan, L.J. Lauhon, T.J. Marks, M.C. Hersam, Carbon nano-materials for electronics, optoelectronics, photovoltaics, and sensing, Chem. Soc. Rev. 42 (7) (2013) 2824—2860.

[22] F.P. Vaccari, S. Baronti, E. Lugato, L. Genesio, S. Castaldi, F. Fornasier, F. Miglietta, Biochar as a strategy to sequester carbon and increase yield in durum wheat, Eur. J. Agron. 34 (4) (2011) 231—238.

[23] M. Saxena, S. Maity, S. Sarkar, Carbon nanoparticles in 'biochar'boost wheat (*Triticum aestivum*) plant growth, RSC Adv. 4 (75) (2014) 39948—39954.

[24] M. Ashfaq, N. Verma, S. Khan, Carbon nanofibers as a micronutrient carrier in plants: efficient translocation and controlled release of Cu nanoparticles, Environ. Sci. Nano 4 (1) (2017) 138—148.

[25] M. Bottini, F. D'Annibale, A. Magrini, F. Cerignoli, Y. Arimura, M.I. Dawson, et al., Quantum dot-doped silica nanoparticles as probes for targeting of T-lymphocytes, Int. J. Nanomed. 2 (2) (2007) 227.

[26] M.C. DeRosa, C. Monreal, M. Schnitzer, R. Walsh, Y. Sultan, Nanotechnology in fertilizers, Nat. Nanotechnol. 5 (2) (2010) 91.

[27] M.N.A. Hasaneen, H.M.M. Abdel-Aziz, D.M.A. El-Bialy, A.M. Omer, Preparation of chitosan nanoparticles for loading with NPK fertilizer, Afr. J. Biotechnol. 13 (31) (2014).

[28] E. Corradini, M.R. De Moura, L.H.C. Mattoso, A preliminary study of the incor-poration of NPK fertilizer into chitosan nanoparticles, Express Polym. Lett. 4 (8) (2010).

[29] H. Lee, S.M. Dellatore, W.M. Miller, P.B. Messersmith, Mussel-inspired surface chemistry for multifunctional coatings, Science 318 (5849) (2007) 426—430.

[30] Y. Lee, H. Lee, Y.B. Kim, J. Kim, T. Hyeon, H. Park, et al., Bioinspired surface immobilization of hyaluronic acid on monodisperse magnetite nanocrystals for targeted cancer imaging, Adv. Mater. 20 (21) (2008) 4154—4157.

[31] R. Ouyang, J. Lei, H. Ju, Surface molecularly imprinted nanowire for protein specific recognition, Chem. Commun. (44) (2008) 5761—5763.

[32] B. Yu, D.A. Wang, Q. Ye, F. Zhou, W. Liu, Robust polydopamine nano/micro-capsules and their loading and release behavior, Chem. Commun. (44) (2009) 6789—6791.

[33] C.J. Ochs, T. Hong, G.K. Such, J. Cui, A. Postma, F. Caruso, Dopamine-mediated continuous assembly of biodegradable capsules, Chem. Mater. 23 (13) (2011) 3141—3143.

[34] X. Jia, Z.Y. Ma, G.X. Zhang, J.M. Hu, Z.Y. Liu, H.Y. Wang, F. Zhou, Polydopamine film coated controlled-release multielement compound fertilizer based on mussel-inspired chemistry, J. Agric. Food Chem. 61 (12) (2013) 2919—2924.

[35] J.J. Joyner, D.V. Kumar, Nanosensors and their applications in food analysis: a review, Int. J. Sci. Technol. 3 (4) (2015) 80.

[36] J. Lu, M. Bowles, How will nanotechnology affect agricultural supply chains? Int. Food Agribus. Manag. Rev. 16 (1030—2016—82815) (2013) 21—42.

[37] E. Omanović-Mikličanin, M. Maksimović, Nanosensors applications in agriculture and food industry, Bull. Chem. Technol. Bosnia Herzegovina 47 (2016) 59—70.

[38] G.N. Rameshaiah, J. Pallavi, S. Shabnam, Nano fertilizers and nano sensors—an attempt for developing smart agriculture, Int. J. Eng. Res. Gen. Sci. 3 (1) (2015) 314—320.

[39] A. Dubey, D.R. Mailapalli, Nanofertilisers, nanopesticides, nanosensors of pest and nanotoxicity in agriculture, in: Sustainable Agriculture Reviews, Springer, Cham, 2016, pp. 307—330.

[40] M. Maksimović, E. Omanović-Mikličanin, Towards green nanotechnology: maximizing benefits and minimizing harm, in: CMBEBIH 2017, Springer, Singapore, 2017, pp. 164—170.

[41] J.A. Dahl, B.L. Maddux, J.E. Hutchison, Toward greener nanosynthesis, Chem. Rev. 107 (6) (2007) 2228—2269.

CHAPTER 11

Polymer formulations for controlled release of fertilizers

Remya V.R., Jesiya Susan George, Sabu Thomas
International and Inter University Centre for Nanoscience and Nanotechnology, Mahatma Gandhi University, Kottayam, Kerala, India

1. Introduction

1.1 Fertilizers

Fertilizer is any type of material which is applied to soil or plant tissues to supply plant nutrients essential to plant growth. Fertilizers can be originated naturally or synthetically (industrially produced) [1]. Since the past 50 years the need for commercial fertilizers has increased steadily. The usage of phosphate fertilizers has increased from 1960 to 2000 [2]. Based on polymers derived from the combination of urea and formaldehyde, controlled-nitrogen-release technologies were produced and commercialized in 1955. The initial products of controllednitrogen-release fertilizers had 60% of the total nitrogen cold-water-insoluble, and less than 15% only the unreacted (quick-release) part. But this problem was solved in 1970s by using methylene urea and having 25% and 60% of the nitrogen as cold-water-insoluble, and 15%—30% of unreacted urea nitrogen. Sulfur-coated urea fertilizers have been developed by National Fertilizer Development Center, and here the principal coating material is sulfur because of its value as a secondary nutrient and its low cost. For developing and enhancing slow-release properties based on degradation of the secondary sealant by mechanical imperfections as well as soil microbes in the sulfur, polymers were used as the sealant in sulfur [3—5]. Fig. 11.1 shows how a fertilizer acts on soil to enhance crop yield [6].

Polymer-coated controlled-release fertilizers (CRFs) have achieved great attention, and this is the newest and most sophisticated method in plant production through a horticultural way. In this technique polymers are coated like a shell over the core nutrients. Mainly the polymer coating is based on the coating thickness and composition of the polymer. Polymer-coated controlled release of fertilizers supply three nutrients rather than conventional one, which is it supplies nitrogen, phosphorous, and potassium

Controlled Release Fertilizers for Sustainable Agriculture
ISBN 978-0-12-819555-0
https://doi.org/10.1016/B978-0-12-819555-0.00011-X

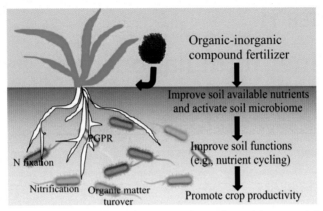

Figure 11.1 Images of fertilizer's action to enhance the crop yield.

along with calcium, sulfur, and other micronutrients. These methods have many advantages like ease of adjusting fertilizers' release rate to many crops, efficiency, and very low ground water pollution. Therefore the discussion of this topic has achieved great importance in this book.

1.2 Mechanism

The main aim of using fertilizers is to enhance the plant growth. Fertilizers work through two ways [7]. In the first one (traditional one), additives provide nutrients to plants. But in the second one, some fertilizers themselves act to enhance the effectiveness of the soil by modifying its water retention and aeration properties. Nitrogen (leaf growth), phosphorus (development of roots, flowers, seeds, fruits), and potassium (movement of water in plants, promotion of flowering and fruiting, strong stem growth) are the major macronutrients provided by fertilizers. They also provide some micronutrients (Cu, Fe, B, Zn, Mn, and Mo) and occasional nutrients (Co, Si, and V). Nitrogen is the most important fertilizer because it is present in the proteins, DNA, and other components. Nitrogen is in fixed form in all nitrogen fertilizers to provide nutrients. Only very small quantities of micronutrients are consumed in plant tissue on the order of parts per million (less than 0.04% DM) [7]. All these elements help to carry out plant metabolism because these elements enable catalysts and their impact far exceeds their weight percentage [8,9]. The mechanism behind the slow release of nitrogen fertilizer is shown in Fig. 11.2 [10]. The mechanism of nitrogen release from urea formaldehyde [UF] slow release fertilizer involves mainly three steps. In the first step, the coating materials

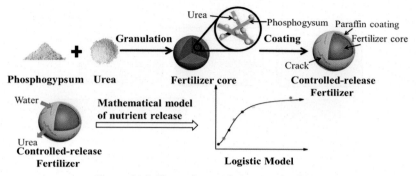

Figure 11.2 Slow release of nitrogen fertilizer.

become swollen by absorbing water from the soil and transfer in to the hydrogel. It will help the diffusion of fertilizer in the core of the gel network. A layer of water between swollen coating and UF granule is created. In the second step, water diffuses into the cross-linked polymer network. This will enhance the dissolution of the soluble part of UF and as a result the soluble part of the fertilizer gets slowly released into the soil. Final step, the soil microorganisms penetrate through the swollen coatings and assemble around the UF granule. Through this way, the insoluble part of nitrogen in UF granule degrades into urea and ammonia and slowly is released into the soil via dynamic exchange [10].

2. Classification of fertilizers

Depending on the enhanced fertilizer use efficiency and minimized leaching, the gradual pattern of nutrient release better meets plants' needs. Polymer-coated controlled-release of fertilizers has attained more attention in the current century than conventional methods. Mainly fertilizers are classified in different ways based on the coating and nutrient element, especially nature of nutrient element, whether to provide single or more nutrients and organic fertilizers or not, is shown in Fig. 11.3.

2.1 Nature of nutrient element

The main nutrient elements present in the fertilizers are nitrogen, phosphorus, and potassium. The nitrogenous chemical fertilizers are urea, calcium, ammonium nitrate, ammonium sulfate, basic calcium nitrate, calcium cyanamide, etc., and these suppy nitrogen to the soil. The main source of nitrogen fertilizer is ammonia, and it is applied directly to the ground. This

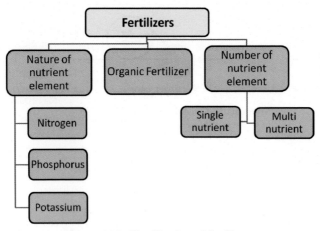

Figure 11.3 Classification of fertilizers.

ammonia acts as a feed stock for all other nitrogen fertilizers. Phosphatic chemical fertilizers supply phosphorus to the soil and triple super phosphate, phosphate of lime, etc., are different examples of the same. These phosphate fertilizers are gained by extraction from minerals containing anionic phosphate. Potassium nitrate, potassium chloride, potassium sulfate, etc., are the different examples of potash chemical fertilizers. Other types of fertilizer are called compound fertilizers which contain N, P, and K and are obtained by the mixing of straight fertilizers. Monoammonium and diammonium phosphates are examples of compound fertilizer [11−16].

2.2 Single nutrient and multinutrient fertilizers

Single and multinutrient classifications are based on whether to provide a single nutrient or two or more nutrients. Ammonium nitrate, urea, etc., are single nutrient nitrogen fertilizers. Single superphosphate and triple super phosphate are examples of single nutrient phosphate fertilizers. In superphosphate fertilizers, 90% is water soluble. Muriate of Potash (MOP) fertilizer is the major type of potassium-based straight fertilizer and contains 95%−99% KCl, and is commonly available as 0-0-60 or 0-0-62 fertilizer. Multinutrient fertilizers contain two or more nutrient elements and are classified as binary fertilizers and NPK fertilizers. NP fertilizers mainly supply nitrogen and phosphorus to the plants. Monoammonium phosphate (MAP) and diammonium phosphate (DAP) are the main NP fertilizers and 85% of MAP and DAP fertilizers are soluble in water. But NPK fertilizers contain nitrogen, phosphorus, and potassium and it is represented as three numbers separated by dashes (e.g., 10-10-10 or 16-4-8). This describes the chemical content of fertilizers [17,18].

2.3 Organic fertilizers

Fertilizers with organic-biologic origin and derived from living or formerly living materials are called organic fertilizers [7]. These fertilizers typically contain both organic materials and acceptable additives such as ground sea shells, nutritive rock powders, etc. Some organic materials from industries like sewage sludge may not be acceptable components of organic gardening and farming. This is due to the residual contaminants to public perception. Peat, coir, etc., are the types of organic soil amendment added to soil and all act similarly because of their limited nutritive inputs [19].

2.4 Coated and uncoated nitrogen-based and polymer-coated multinutrient fertilizers

Chemically bound urea is the oldest CRF in which release rate is determined by particle size, available water, and microbial decomposition. Urea form and isobutylidene diurea (IBDU) are the examples of these types of uncoated nitrogen-based fertilizer. The first CRF-type nitrogen release fertilizer is sulfur-coated urea. Here the nitrogen release is mainly based on the thickness of sulfur coating. The polymer-coated CRF fertilizers are mainly composed of nutrient core and polymer shell. This is the newer and technically more applicable fertilizer. Osmocote, Apex, multicote, nutricote, and diffusion type fertilizers are examples of polymer-coated fertilizers.

3. Polymer formulations for controlled release of fertilizers

Polymer-coated controlled-release fertilizers are the most commonly used and recently trending, and effective in forest, conservation, and native plant nurseries rather than other conventional fertilizers. The controlled release of these fertilizers is based on the release rate, that is, these fertilizers release their nutrients over periods from 3 to 18 months based on coating and temperature of the medium. For growers, polymer-coated CRF (PCRF) have more advantages including ease of adjusting fertilizer rate for many crops, less ground water pollution, and better use efficiency. Therefore nowadays, PCRF has been used in bareroot nurseries to produce quality seedlings with less expense. CRFs are the fertilizer granules incorporated with some carrier molecules and thereby control the release of fertilizers; hence improving the nutrient supply and reducing environmental and health hazards. With CRFs, the principal method is to cover a conventional soluble fertilizer with a protective coating (encapsulation) of a water-insoluble, semipermeable, or impermeable-with-pores material. This controls water penetration and thus the rate of dissolution and ideally synchronizes nutrient release with the plants' needs. The thickness of the coating over the fertilizer enables it to release over an extended period of time and thereby controls the pattern of release and nutrient efficiency, and can eliminate the constant supply of large quantity of fertilizers in the fields. Sulfur-coated urea (SCU) is the first commercialized and studied slow release fertilizer, in which urea granules are uniformly coated with sulfur and wax to seal the cracks and pores. The efficiency of the product depends on size of the urea prill or granules and the thickness of coating. The release of the drug can be accomplished by either of the these two mechanisms: one is the biodegradation of wax and/or sulfur; second one is the formation of cracks and pores created due to the conversion of amorphous polymer to crystalline polymer; in addition to these environmental factors such as moisture and temperature also affect the release rate. However SCU has some disadvantages too; incorporation of sulfur makes the coating less resistant to abrasion and sometimes sulfur may be oxidized into sulfuric acid, resulting in soil acidity [20–22]. Fig. 11.4 shows the classification of CRFs [35].

3.1 Polymer-based CRF

Plastics as well as various elastomers are used for the preparation of CRFs. Till date several water-insoluble polymeric coatings are used in CRF formulation, such as polystyrene, polyacrylonitrile, polyethylene, Kraft pine lignin, and polyacrylamide. The release course consists of three distinct stages (1) the initial stage during which almost no release is observed (lag period), (2) a constant-release stage, and (3) a stage of gradual decay of release rate. It was assumed that the duration of the lag period was linked to the time needed for the internal voids of a coated granule to fill with a critical amount of water and thus induce good contact of the solution with the inner side of coating, after which a steady state between water penetrating into the granule and nutrients leaving it is attained. The stage of linear release lasts as long as there is solid fertilizer in the granule and thus a constant gradient between the granule and medium solutions is, practically, maintained. Sulfur coatings for fertilizers were frequently used compared to polymer-based coatings since the sulfur coating is easy to break than a polymer coating. The release rate of nutrients from a polymer-coated fertilizer is affected by factors such as coating thickness, rate of diffusion, and soil temperature [23–25].

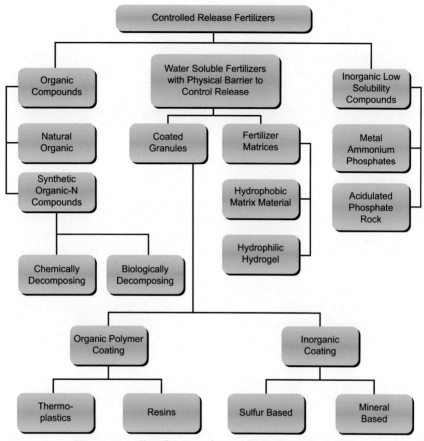

Figure 11.4 Classification of controlled-release fertilizers.

3.2 Starch and its derivatives

Starch is an organic material produced by plants; moreover, it is a polysaccharide consisting of glucose moieties linked together by α 1, 4 and/or 1, 6 linkages. The presence of abundant hydroxyl groups in starch creates many options for its modification; in addition to modification, blends of starch with other polymers are also used. Hu et al. reported starch/polyvinyl alcohol (PVA) ecofriendly water-soluble coatings for fertilizers. They mixed various amounts of starch and PVA in distilled water along with glycerol and butanol, former will act as a plasticizer and later is used to avoid frothing. The water absorbency results of the prepared films show that water permeability, absorbency, and NH_4 permeability of the films increases by increasing the amount of polyvinyl alcohol content, because

PVA has a lot of surface hydroxyl groups than starch. Hence there will be an increase in the hydrophobicity, polarity, and crystallinity by increasing PVA content in the starch/PVA films. Formaldehyde cross-linking of these films causes intermolecular and intermolecular $-OH$ linkages between starch and PVA, thereby increasing hydrophobicity after cross-linking with formaldehyde. Recently Jyothi et al. [23] also reported the starch/PAN coated urea fertilizers with enhanced slow release and water absorbance. They have synthesized starch/PAN copolymers by free radical polymerization of pregelatinized starch and acrylonitrile using ceric ammonium nitrate as initiator. The prepared copolymer was applied as a coating for urea fertilizer using a coating machine. Fig. 11.5 represents the microscopic images of urea particles coated with copolymer coating. Coated urea exhibits a better controlled release than the uncoated one and the rate of release depends on the grafting percentage of the copolymer. Presence of hydrophobic acrylonitrile groups in the coating formulation drastically reduces the penetration of water molecules into the urea and thereby reduces the dissolution and subsequent release into the soil [26,27]. Similarly several works are reported on starch/chitosan, starch/PLLA, starch/polyvinyl acetate coatings for the controlled release of fertilizers [28–31].

3.3 Polyurethane

Polyurethanes are a class of polymers that contain a characteristic urethane linkage in them ($-NHCOO-$). They are generally formed by the reaction between polyols (diol/triol) and isocyanate (diisocyanate/polyisocyanate). They are mainly used in the areas such as adhesive, coatings, cushions, shoes, and so on. They are also used as a good candidate for controlled release of fertilizers. Ma et al. prepared bio-based polyurethane from cotton straw; they have further modified it with siloxane and polyether to increase

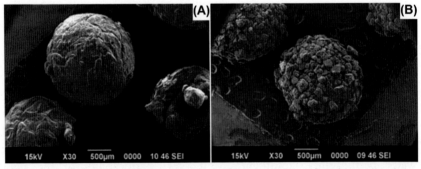

Figure 11.5 SEM image of urea particle without coating and with coating (open access).

the hydrophobicity as well as to reduce micropore size and number. The cotton straws were liquefied using a reflux condenser. Bio-based polyurethane-coated fertilizer was obtained by rotating heating drum machine along with the native fertilizer in a rotator using liquefied cotton straw and polyaryl polymethylene isocyanate; however, for the preparation of siloxane modified PU, same procedure was used by adding extra siloxane and polyether. Siloxane-modified BPCF and DBPCF exhibit good N release efficiency. Siloxane modification enhances the hydrophobicity and thereby reduces the water intake into the fertilizer, whereas the polyether modification increases the N release capacity. However the synergistic effects of these two modifications enhance the overall performance of the fertilizer [32]. Liu et al. reported biodegradable, environment friendly palm oil-based polyurethane coatings for urea. Epoxidation of the palm oil was carried out using peracid, later polyols were prepared from peroxidized palm oil using methanol via ring opening of formed epoxy ring. Prepared bio-based coatings were coated on urea prills by rotary drum machine; acrylonitrile modification on this palm oil PU increases the hydrophobicity and swelling capacity [33].

3.4 Polysulfone

Polysufone (PS) belongs to the class of high performance polymer; they are relatively well known for their thermal stability and high temperature resistance. Tomaszewska et al. reported polysulfone/starch coatings for NPK fertilizer. For the preparation of coating formulation solid polysulfone was dissolved in N,N' dimetyloformamide; once PS dissolved homogenously, starch was added into that in various compositions. The final coating formulation was prepared by phase inversion technique. They found that increased amount of starch in the composition increases the hydrophobicity and thereby enhances the water diffusion rate and fertilizer release [34]. Fig. 11.6 represents the diffusion mechanism of the controlled release of fertilizer, water molecules in the field diffuse into the fertilizer through coating, condenses on the solid fertilizer, and results in the slow release of fertilizer [35].

Osmocote (polymer resin-coated), apex (polyon), multicote-nutricote (thermoplastic-coated) are the different types of other polymer-coated CRFs. These are differentiated by varying nutrient element, release pattern, and longevity. Therefore polymer formulations of controlled release of fertilizers have attained much more importance for uniform and healthy plant growth [24].

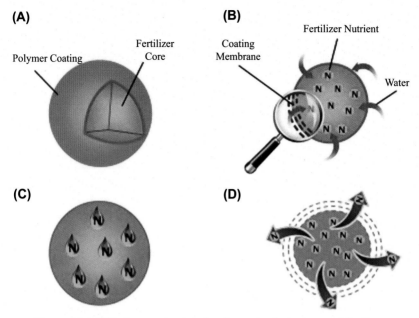

Figure 11.6 Diffusion mechanism for the controlled release of fertilizer.

4. Conclusions

The development of more novel, stable, useful, and sustainable agro-ecosystems like fertilizers for improving food production has caused wide public concern in recent years. In this chapter, we discussed the mechanism and classification of fertilizers and detailed the polymer formulations for controlled release of fertilizers [PCRF] and various polymer-coated fertilizers and their action to soil. Controlled release based on polymer/super-absorbent materials offers promising potential in terms of extended controlled release and water retention. However, the complexity of processing, elevated costs, and the non–environmentally-friendly side effects of some materials prevent industrial-scale production. Therefore, recently some research works were reported with regard to the production of controlled release of fertilizers with starch-, lignin-, and cellulose-based coating materials, which are relatively cheaper, biodegradable, and renewable. Anyway, this chapter provides knowledge of fertilizers and the importance of polymer formulations for controlled release of fertilizers; it also discusses the cheap, biodegradable, and environmental friendly fertilizers and their usage and mechanism to deliver nutrients to soil for enhancing crop yield.

References

[1] H.W. Scherer, Fertilizers in Ullmann's Encyclopedia of Industrial Chemistry, Wiley-VCH, Weinheim, 2000.

[2] Commercial Fertilizers Increase Crop Yields [1]. (Accessed 9 April 2012).

[3] C.P. Vance, C. Uhde-Stone, D.L. Allan, Phosphorus acquisition and use: critical adaptations by plants for securing a nonrenewable resource, New Phytol. 157 (3) (2003) 423–447.

[4] Mergers in the Fertiliser Industry. The Economist, February 18 , 2010. (Accessed 21 February 2010).

[5] J.B. Sartain, University of Florida, Food for Turf: Slow-Release Nitrogen, Grounds Maintenance, 2011.

[6] J. Zhao, T. Ni, J. Li, Q. Lu, Z. Fang, Q. Huang, Q. Shen, Effects of organic—inorganic compound fertilizer with reduced chemical fertilizer application on crop yields, soil biological activity and bacterial community structure in a rice—wheat cropping system, Appl. Soil Ecol. 99 (2016) 1–12.

[7] H. Dittmar, M. Drach, R. Vosskamp, M.E. Trenkel, R. Gutser, G. Steffens, Fertilizers, 2 Types. Ullmann's Encyclopedia of Industrial Chemistry, 2009.

[8] AESL Plant Analysis Handbook — Nutrient Content of Plant Aesl.ces.uga.edu. Retrieved: 11 September 2015.

[9] H.A. Mills, J.B. Jones Jr., Plant Analysis Handbook II: A Practical Sampling, Preparation, Analysis, and Interpretation Guide, 1996.

[10] X. Yu, B. Li, Release mechanism of a novel slow-release nitrogen fertilizer, Particuology 45 (2019) 124–130.

[11] V. Smil, Enriching the Earth, Massachusetts Institute of Technology, 2004, p. 135.

[12] J.B. Jones Jr., Plant Nutrition and Soil Fertility Manual, CRC press, 2012.

[13] Supplemental Technical Report for Sodium Nitrate (Crops) www.ams.usda.gov. Retrieved: 6 July 2014.

[14] V. Gowariker, V.N. Krishnamurthy, S. Gowariker, M. Dhanorkar, K. Paranjape, The Fertilizer Encyclopedia, John Wiley & Sons, 2009.

[15] Caliche Ore. www.sqm.com. Archived from the original on 14 July 2014. Retrieved: 6 July 2014.

[16] F. Brentrup, A. Hoxha, B. Christensen, Carbon footprint analysis of mineral fertilizer production in Europe and other world regions, In: Conference Paper, The 10th International Conference on Life Cycle Assessment of Food (LCA Food 2016).

[17] Label Requirements of Specialty and Other Bagged Fertilizers. Michigan Department of Agriculture and Rural Development. Retrieved: 14 March 2013.

[18] Summary of State Fertilizer Laws (PDF). EPA. Retrieved: 14 March 2013.

[19] R.J. Haynes, R. Naidu, Influence of lime, fertilizer and manure applications on soil organic matter content and soil physical conditions: a review, Nutrient Cycl. Agroecosyst. 51 (2) (1998) 123–137.

[20] S.E. Allen, C.M. Hunt, G.L. Terman, Nitrogen release from sulfur-coated urea, as affected by coating weight, placement and temperature 1, Agron. J. 63 (4) (1971) 529–533.

[21] C.F. Yamamoto, E.I. Pereira, L.H. Mattoso, T. Matsunaka, C. Ribeiro, Slow release fertilizers based on urea/urea—formaldehyde polymer nanocomposites, Chem. Eng. J. 287 (2016) 390–397.

[22] J. Chen, S. Lü, Z. Zhang, X. Zhao, X. Li, P. Ning, M. Liu, Environmentally friendly fertilizers: a review of materials used and their effects on the environment, Sci. Total Environ. 613 (2018) 829–839.

[23] K. Zhong, Z.T. Lin, X.L. Zheng, G.B. Jiang, Y.S. Fang, X.Y. Mao, Z.W. Liao, Starch derivative-based superabsorbent with integration of water-retaining and controlled-release fertilizers, Carbohydr. Polym. 92 (2) (2013) 1367—1376.

[24] T.D. Landis, R.K. Dumroese, Using polymer-coated controlled-release fertilizers in the nursery and after outplanting, Forest Nursery Notes. Winter: 5-12, (2009) 5—12.

[25] S.I. Sempeho, H.T. Kim, E. Mubofu, A. Hilonga, Meticulous overview on the controlled release fertilizers, Adv. Chem. (2014).

[26] K. Zhong, Z.T. Lin, X.L. Zheng, G.B. Jiang, Y.S. Fang, X.Y. Mao, Z.W. Liao, Starch derivative-based superabsorbent with integration of water-retaining and controlled-release fertilizers, Carbohydr. Polym. 92 (2) (2013) 1367—1376.

[27] X. Han, S. Chen, X. Hu, Controlled-release fertilizer encapsulated by starch/polyvinyl alcohol coating, Desalination 240 (1—3) (2009) 21—26.

[28] J.J. Perez, N.J. Francois, Chitosan-starch beads prepared by ionotropic gelation as potential matrices for controlled release of fertilizers, Carbohydr. Polym. 148 (2016) 134—142.

[29] L. Chen, Z. Xie, X. Zhuang, X. Chen, X. Jing, Controlled release of urea encapsulated by starch-g-poly (L-lactide), Carbohydr. Polym. 72 (2) (2008) 342—348.

[30] Y. Niu, H. Li, Controlled release of urea encapsulated by starch-g-poly (vinyl acetate), Ind. Eng. Chem. Res. 51 (38) (2012) 12173—12177.

[31] A.N. Jyothi, S.S. Pillai, M. Aravind, S.A. Salim, S.J. Kuzhivilayil, Cassava starch-graft-poly (acrylonitrile)-coated urea fertilizer with sustained release and water retention properties, Adv. Polym. Technol. 37 (7) (2018) 2687—2694.

[32] X. Ma, J. Chen, Y. Yang, X. Su, S. Zhang, B. Gao, Y.C. Li, Siloxane and polyether dual modification improves hydrophobicity and interpenetrating polymer network of bio-polymer for coated fertilizers with enhanced slow release characteristics, Chem. Eng. J. 350 (2018) 1125—1134.

[33] J. Liu, Y. Yang, B. Gao, Y.C. Li, J. Xie, Bio-based elastic polyurethane for controlled-release urea fertilizer: fabrication, properties, swelling and nitrogen release characteristics, J. Clean. Prod. 209 (2019) 528—537.

[34] M. Tomaszewska, A. Jarosiewicz, Use of polysulfone in controlled-release NPK fertilizer formulations, J. Agric. Food Chem. 50 (16) (2002) 4634—4639.

[35] B. Azeem, K. KuShaari, Z.B. Man, A. Basit, T.H. Thanh, Review on materials & methods to produce controlled release coated urea fertilizer, J. Contr. Release 181 (2014) 11—21.

CHAPTER 12

Chemistry and toxicology behind chemical fertilizers

Stalin Nadarajan[1], Surya Sukumaran[2]
[1]Institute of Plant Science, ARO Volcani Center, Rishon Lezion, Israel; [2]School of Pure and Applied Physics, Mahatma Gandhi University, Kottayam, Kerala, India

1. Introduction

The growth and progression of economic development in the early stages of India was predominantly dependent upon the progress of agricultural sector. Such an agricultural productivity can only be achieved either by bringing more land area under cultivation or via increasing the productivity [18]. Nowadays, the expansion of global population has resulted in lower carrying capacity of land [18]. The prime role of agriculture is not restricted to its contribution toward national income, but also to meet the extended food security of our nation. Hence, the increase in the rate of intensive farming, in turn adversely affected the soil fertility and eventually resulted in localized ecological disasters [5,53]. Later, it was realized that the pattern of production enhancement would have to rest profoundly on increased yield. The existing farmlands were optimized by adopting new strategies of farming by green revolution to maintain sustainable agricultural productivity and one among them was the judicious use of chemical fertilizers [10,18]. The green revolution helped to overcome food scarcity but it substantially increased the dependence on synthetic fertilizers in agriculture [10]. Fertilizer is one among the key elements maintaining the tempo of agricultural production by offering better quality and surplus yield for exporting and future purposes [18]. It enriches soils with specific micro/macro nutrients and are compounds of synthetic origin offering high concentration of nutrients to soil. When applied, fertilizers get broken down into chemical components and are then absorbed through the root system as a part of plant's nourishment. They are extremely attractive because of their costeffectivenes, easy availability, high nutrient concentration, convenient transportation, and ease of applications [11]. In the pursuit of economic growth and food production, increasing amounts of

Controlled Release Fertilizers for Sustainable Agriculture
ISBN 978-0-12-819555-0
https://doi.org/10.1016/B978-0-12-819555-0.00012-1

chemical fertilizers have been applied in agro ecosystems all over the world [50,56].

Most of the inorganic fertilizers influencing the growth and development of plants are NPK fertilizers that are rich in macronutrients nitrogen (N), phosphorus (P), and potassium (K). Moreover fertilizers also supplement minor nutrients like calcium, sulfur, magnesium to enhance the soil fertility. The fertilizer industries are the major contributors of outdoor terrestrial natural radionuclides (^{238}U, ^{232}Th, and ^{210}Po) and heavy metals (Hg, Cd, As, Pb, Cu, Ni, and Cu) as potential sources [22,57]. The factory workers associated with the production of fertilizers and farmers involved in direct application of fertilizers are exposed to ionizing radiations from these radionuclides. The long–term exposure of such radiations released from fertilizers has a potential to trigger cancers in humans [57]; hence, international awareness about such radiation hazards should be taken into consideration.

Over the past decades, chemical fertilization increased exponentially throughout the world causing serious environmental pollution and health hazards. This leads to the deterioration of soils' natural properties and fertility, accumulation of heavy metals in plant tissues, compromising the nutritional value of fruits and other crops [55]. Although chemical/synthetic fertilizers have been claimed to be the most important contributor to the world's agricultural productivity [58], their negative effects on environment and humans limit its usage in sustainable agricultural systems [47]. Hence, it is mandatory to gain a detailed knowledge about the chemistry and journey of agropollutants from soils to ecosystems and from crops to its dependent food chains via absorption from soil. These toxic agropollutants persist in the soil and leach into water sources causing serious environmental issues like increasing the rate of eutrophication, loss of biodiversity, ozone layer depletion, global warming, ground water pollution, and threatening of the future food security, thus, leading to health problems [10]. Going ahead with the awareness of sustainable agriculture, it is essential to explore the link between the chemistry and toxicity effects of commonly used chemical fertilizers.

2. Types of fertilizers

Fertilizers are classified based on the types and the form of chemical compound they contain. Fig.12.1 depicts the different types of fertilizers used for plant growth and development.

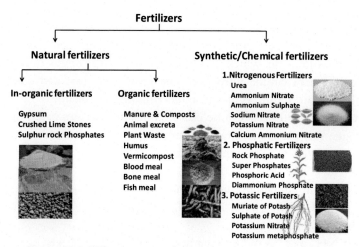

Figure 12.1 Different types of organic and inorganic fertilizers.

3. Chemistry of chemical fertilizer

Chemical fertilizers (CFs) are manufactured using the different proportions of major and minor chemical nutrients. Most CFs contain significant amount or percentage of the three major nutrients N, P, and K. Due to different usage purposes regarding planting medium and soil nutrient conditions, CFs are produced with this in mind. There are different types of CFs, all manufactured for application in different soil conditions for a variety of crops and plants. Most of the CF products are labeled with three numbers, such as 10−20−30. These numbers refer to the N, P, K content present in the CF. A complete chemical fertilizer will have all three elements in significant percentages. Some CFs contain urea and ammonium along with these three as a common addition. Occasionally, there can be a smaller percentage of nutrients like copper, zinc, iron, manganese, and sulfur in the fertilizer. These nutrient elements are also necessary for healthy growth of plants. Chemical fertilizer industries are synthesizing CFs through chemical processes using naturally occurring deposits, while chemically altering them. Generally, fertilizer industry is involved in the production of primary plant nutrients suitable for application in the soil. This chapter discusses the chemistry and synthesis process of three major types of CFs and their positive and negative effects on the soil health conditions.

3.1 Nitrogenous fertilizers

Globally, application of nitrogenous fertilizer has increased rapidly and it is expected to increase by four to fivefold by 2050 with two-third of that application in developing countries [62]. Nitrogenous fertilizers are those fertilizers which contain N as the chief component in their final product. In nitrogenous fertilizer, N is present as ammoniacal nitrogen such as ammonium chloride, ammonium sulfate; nitrate nitrogen like calcium ammonium nitrate in which both ammoniacal and nitrate nitrogen are present; and urea (amide nitrogen). The most important and commonly used nitrogenous fertilizers are urea and ammonium sulfate.

Nitrogenous fertilizers can be classified into four classes based on forms of nitrogen (N) present in straight nitrogenous fertilizers:

1. Nitrate nitrogen-containing fertilizers (NO_3-N):
   e.g., $NaNO_3$—16% N; $Ca(NO_3)_2$—15.5% N
2. Ammonium containing nitrogenous fertilizers (NH_4-N):
   e.g., $(NH_4)_2SO_4$—20% N; NH_4Cl—24% to 26% N, anhydrous ammonia, 82% N
3. Both NH_4 and NO_3-N containing nitrogenous fertilizers:
   e.g., ammonium nitrate (NH_4NO_3)—33% to 34% N, calcium ammonium nitrate—20% N
4. Amide fertilizers—it is the organic form of N-containing fertilizers:
   e.g., Urea [$CO(NH_2)_2$]—46% N, calcium cyanamide ($CaCN_2$)—21% N.

3.1.1 Ammonium sulfate [$(NH_4)_2SO_4$]

Ammonium sulfate fertilizer is a stable white to yellowish-gray crystalline salt that is soluble (750 g/L) in water. It contains about 21% N and 24% S. This fertilizer is mainly used for alkaline soil. A small amount of acidity forms in the soil when ammonium is released and it leads to lowering the pH balance of the soil, while contributing to essential nitrogen for plant growth and development.

3.1.1.1 Synthesis

When nitrogen and hydrogen in the ratio of 1:3 are passed over heated (500°C) platinized asbestos or powdered iron at a pressure of 250 atm, ammonia is formed which is further passed through a suspension of gypsum in water through which carbon dioxide is also passed. The resulting calcium carbonate is removed by filtration. When aqueous solution of ammonium sulfate is evaporated, crystals of ammonium sulfate are formed, which are dried and bagged.

$$N_2 + H_2O \xrightarrow[250atm]{500°C} 2NH_3$$

$$2NH_3 + CO_2 + H_2O \longrightarrow (NH_4)_2CO_3 \xrightarrow{CaSO_4} (NH_4)SO_4 + CaCO_3$$

Ammonium sulfate has excellent storage properties. The solution form of ammonium sulfate is acidic to litmus. In this fertilizer, N is present in cationic form. This form of N can be retained by soil colloids; hence, loss of nitrogen through leaching is less. In addition, it is less hygroscopic and therefore poses less problem when handling. The application of heavy dose of ammonium sulfate to alkaline soils may lead to the loss of N through volatilization.

3.1.1.2 Reactions in soil

When ammonium sulfate is applied to upland soil, the first reaction is the cation–exchange, in which ammonium (NH_4^+) displaces some other cations, usually calcium, on the exchange complex of the soil. The resulting calcium sulfate [$CaSO_4$] being soluble in water is partly taken up by plants but mostly leached through the soil during heavy rains. In this way, the loss of soil calcium (Ca^{2+}) takes place due to application of ammonium sulfate fertilizer. The ammonium (NH_4^+) on the soil colloid may be taken up directly by the plants or it is nitrified. The biological process of oxidation of NH_4^+ to NO_3^- is known as *nitrification*. It is a two-step process in which the ammonium (NH_4^+) is first converted to nitrite (NO_2^-) by nitrosomonas and then to nitrates (NO_3^-) by nitrobacter. This reaction indicates that hydrogen (H^+) ions are formed during nitrification which later result in the acidification of the soil and subsequently lower the soil pH, thereby causing soil acidity. This is the reason why ammonium fertilizers are called *physiologically acidic fertilizers*. If a soil contains huge amounts of free $CaCO_3$, then the loss of Ca^{2+} will fall upon the $CaCO_3$ and not on the exchangeable bases. Under such conditions, there will be no development of acidity due to ammonium sulfate fertilizer application.

$$Ca\ \boxed{Soil\ colloids} + (NH_4)_2SO_4 \longrightarrow \overset{NH_4}{\underset{NH_4}{}}\boxed{Soil\ colloids} + CaSO_4 \downarrow$$
$$\text{Leaching}$$

$$Ca\ \boxed{Soil\ colloids} + 2HNO_3 \longrightarrow \overset{H}{\underset{H}{}}\boxed{Soil\ colloids} + Ca(NO_3)_2 \downarrow$$
$$\text{Leaching}$$

$$2NH_4^- + 3O_2 \xrightarrow{\text{Nitrosomonas}} 2NO_2^- + 2H_2O + 4H^+$$

$$2NO_2^- + O_2 \xrightarrow{\text{Nitrobacter}} 2NO_3^-$$

3.1.2 Urea [(NH₂)₂CO]

Urea is the world's most common and widely used nitrogen fertilizer in the agricultural sector for improved plant growth and yield. This chemical fertilizer has been used as a ladder to huge success in agro revolution among farmers. Urea is an organic solid nitrogen fertilizer with an NPK ratio of 46–0–0. It is a highly soluble (1079 g/L at 20°C) compound and easily gets along with water. The chemical structure of urea helps it to be soluble enough, as the hydrogen bonds with water molecules each form two bonds with oxygen. The planar structure further enhances its astonishing molecular point symmetry. The chemical formula of urea is very close in composition to the organic formulation of the urea, which also benefits crops due to its mass use for feeding major cultivated crops all over the world. Even though urea is naturally produced in humans and animals, commercial synthetic urea is manufactured with the help of anhydrous ammonia and carbon dioxide. Urea is known worldwide by a variety of names like carbamide resin, isourea, carbonyl diamide, and carbonyldiamine.

3.1.2.1 Synthesis

Urea is manufactured by reacting anhydrous ammonia (NH_3) and carbon dioxide (CO_2) gas under intense pressure (~ 400 atm) and temperature (160–200°C). The intermediate product (ammonium carbamate) is unstable; it decomposes and urea is recovered. Urea crystals are formed once concentrated to 99% and sprayed into chamber where urea crystals are formed. During the concentration of the solution or crystallization or preparation of granulated urea at high temperature, some amount of biuret may be formed by condensation reaction. This biuret is toxic to plant; hence, the biuret concentration in urea should not exceed 1.5%. When urea will be applied as foliar spray on crop canopy, the biuret concentration should not exceed 0.25%. Pure solution of urea is slowly hydrolyzed into NH_3 and CO_2 in the absence of bacteria or enzyme. However, this reaction is accelerated at high temperature in combination with the addition of alkali or acid and also by adding bacteria and urease enzyme. Urea reacts with HNO_2 releasing elemental N and CO_2. At 86°F, urea absorbs moisture from the atmosphere if the relative humidity is 72% or above and releases moisture to the atmosphere if the humidity is below 72%. Urea is less acidic compared to ammonium sulfate.

$$NH_3 + CO_2 \underset{400atm}{\overset{160\text{-}200°C}{\rightleftharpoons}} NH_2COONH_4$$

$$\downarrow \text{Unstable Ammonium Carbamate}$$

$$NH_2CONH_2 + H_2O$$
$$\text{Urea}$$

During the concentration of the solution for preparation of granulated urea at high temperature, some amount of biuret may be formed by condensation reaction.

$$NH_2CONH_2 + NH_2CONH_2 \xrightarrow{\text{Condensation reaction}} NH_2-CO-NH-CO-NH_2 + NH_3 \uparrow$$
$$\text{[Biuret]}$$

3.1.2.2 Reactions in soil

Urea in soil undergoes enzymatic hydrolysis to form an unstable ammonium carbamate compound and then converted into ammonia and carbon dioxide.

$$NH_2CONH_2 + H_2O \rightleftharpoons NH_2COONH_4$$
$$\downarrow \text{Unstable Ammonium Carbamate}$$
$$NH_3 + CO_2$$

This NH_3 is converted to NH_4^+ ions by accepting one proton (H^+) from proton donor and subsequently forms NH_4OH or any other NH compound depending upon the nature of the donor. After formation of NH_4 compound, its behavior in the soil and effect on the soil property will be similar to that of any other nitrogenous fertilizers as ammoniacal form.

In neutral, slightly alkaline, and calcareous soils, the chances of formation of nitrites (NO_2^-) are more than in acid soils. When urea is applied in relatively high amounts, nitrites accumulate even in acid soils, because the pH of the soil increases during the hydrolysis of urea. In high concentration of urea, there is a combined toxic effect of NH_3 and NO_2 (nitrite). Due to higher concentration of NH_3, the conversion of NH_4 to NO_2 will increase. However, due to lower population of nitrobacter, the conversion of NO_2 to NO_3 will be inhibited.

$$NH_2CONH_2 + H_2O \xrightarrow{\text{Urease}} (NH_4)_2CO_3 + 3O_2 \xrightarrow[\text{oxidation}]{\text{Microbial}} 2NHO_2 + 3H_2O + CO_2$$
$$2HNO_2 + 3H_2O + CO_2 \xrightarrow[\text{oxidation}]{\text{Microbial}} 2HNO_3$$

3.2 Phosphatic fertilizers

Phosphate fertilizers are widely used in agriculture all over the world [27], since phosphorus (P) is the second essential macronutrient with several functions in the plant system. P is a vital constituent for normal plant growth and development and is among the major drivers for global crop productivity [60]. It helps a plant to convert other nutrients into useable building blocks required for normal plant growth and development [9]. P also plays a vital role in the array of metabolic and biochemical processes within the plant system such as photosynthesis; respiration; energy metabolism; synthesis of nucleic acids, lipids, and proteinoid membranes [38]. It has significant functions in seed germination, seedlings development, root growth, flowering, and fruit formation [2,25,68]. Phosphorus deficit is a major nutritional disorder limiting crop productivity on many low P input agro-ecosystems around the globe [16].

This is a nonrenewable natural resource present in all types of rocks and soils. Phosphate is found in different forms such as dolomite, apatite, iron oxide minerals, calcite, and clay minerals. In commercial production, P fertilizers are applied to increase yields and crop quality. Crops are frequently supplied with inorganic phosphate fertilizers to overcome native soil P deficiency and to improve crop yields [31]. The average world P-use efficiency for cereal production is seriously low and estimated to be about 16% only [19]. Enhancing P-use efficiency is imperative to sustain P sources and crop productivity because of the nonrenewable nature of phosphate rock reserves, geopolitics among countries, and low use efficiency of costly phosphatic fertilizers [16,40,43].

3.2.1 Synthesis

Ammonium phosphate fertilizers first became available in the 1960s. Rock phosphate is one of the basic raw materials required in the manufacture of phosphatic fertilizers such as single superphosphate (SSP), diammonium phosphate (DAP), nitrophosphates, etc. Sulfuric acid is mostly used for the manufacture of phosphate fertilizer. Phosphoric acid is produced by reacting phosphate rock with sulfuric acid, i.e., reaction of the phosphoric acid with ammonia to make monoammonium phosphate $(NH_4H_2PO_4(s))$ and diammonium phosphate $(NH_4)_2H_2PO_4(s)$. These are often referred to as MAP and DAP [35]. Commercial rock phosphate occurs in nature as deposits of apatites (bearing minerals) along with other accessory minerals such as quartz, silicates, carbonates, sulfates, sesquioxides, etc. Four types of rock phosphate minerals are carbonateapatite $[3Ca_3(PO4)_2.CaCO_3]$, fluorapatite $[3Ca_3 (PO4)_2.CaF_2]$, hydroxyapatite $[3Ca_3(PO4)_2.Ca(OH)_2]$, and sulphoapatite $[3Ca_3 (PO4)_2.CaSO_4]$. Due to their well-developed crystalline formation property, the apatites of igneous and metamorphic origin are generally regarded as less reactive.

3.2.2 Diammonium phosphate (DAP)
3.2.2.1 Chemical properties

DAP [$(NH4)_2HPO_4$] is the world's most commonly used phosphorus fertilizer by growers today. It is made from two common macronutrients (phosphate and nitrogen at $18N-46P_2O_5-0K_2O$ content) and it is popular because of its relatively highest concentration of phosphate and nitrogen content coupled with its excellent physical properties.

DAP fertilizer is an excellent source of P and N for plant nutrition. It is highly soluble (588 g/L at 20°C) and thus dissolves quickly in soil to release plant-available phosphate and ammonium. A notable property of DAP is the alkaline pH (7.5−8) that develops around the dissolving granule. As ammonium is released from dissolving DAP granules, volatile ammonia can be harmful to seedlings and plant roots in immediate proximity. This potential damage is more common when the soil pH is greater than 7, a condition that commonly exists around the dissolving DAP granule. To prevent the possibility of seedling damage, care should be taken to avoid placing high concentrations of DAP near germinating seeds. The ammonium present in DAP is an excellent N source and will be gradually converted to nitrate by soil bacteria, resulting in a subsequent drop in pH. Therefore, the rise in soil pH surrounding DAP granules is a temporary effect. This initial rise in soil pH neighboring DAP can influence the microsite reactions of phosphate and soil organic matter.

3.2.3 Monoammonium phosphate (MAP)

MAP fertilizer is basically the same as DAP, but has a lower concentration of nitrogen with the highest P content of any other solid fertilizer at $11 N-52P_2O_5-0K_2O$. It is completely water-soluble (370 g/L at 20°C), has granular form, and is used in the formulation of suspension fertilizers. Being present in granular form, it mixes well and often serves as a constituent in bulk-blended fertilizers. Since it contains less ammonia-nitrogen content compared to DAP, it is considered an ideal starting fertilizer with reduced seed damage potential.

Plants absorb phosphorus in the form of $H_2PO_4^-$ and HPO_4^{2-}. Three replaceable ions of H_3PO_4 combine with Ca^{2+} to form three different combined salts of calcium and phosphorus resulting in different classes of phosphatic fertilizers. Phosphatic fertilizers can again be classified into three groups on the basis of the forms in which orthophosphoric acid (H_3PO_4) combines with calcium (Ca^{2+}):

1. Water-soluble monocalcium phosphate [$Ca(H_2PO_4)_2$]

Single superphosphate (SSP)—16%—18% P_2O_5 or 6.88%—7.74% P
Double superphosphate (DSP)—32% P_2O_5 or 13.76% P
Triple superphosphate (TSP)—46%—48% P_2O_5 or 19.78%—20.64% P
They contain water-soluble P and can be easily available to plants as H_2PO_4 ions. However, within a very short time, this class of fertilizers is converted into insoluble phosphorus when it is applied to the soil. The magnitude of such reaction with the soil depends on the nature and properties of soil.

2. Citric acid-soluble, dicalcium phosphate [$Ca_2H_2(PO_4)_2$, or $CaHPO_4$]
   Basic slag, silicates of lime, 14%—18% P_2O_5 or 6.02%—7.74% P
   Dicalcium phosphate, 34%—39% P_2O_5 or 14.62%—16.77%
   This class of fertilizers is not readily soluble in water and hence not readily available to plants but suitable for acidic soils.

3. Phosphatic fertilizers not soluble in water or citric acid, tricalcium phosphate [$Ca_3(PO_4)_2$].
   Rock phosphate, 20%—40% P_2O_5 or 8.6%—17.2% P
   Raw bone meal, 20%—25% P_2O_5 or 8.6%—10.75% P
   Steamed bone meal, 22% P_2O_5 or 9.46% P
   This class of fertilizers is suitable for strongly acidic soils or organic soils.
In India, most important phosphatic fertilizers used by the farmers are superphosphates and rock phosphates.

3.2.4 Reactions in soils
Effectiveness of phosphatic fertilizers is determined by the properties of both the P salt and the soil being fertilized, as well as the reactions which occur between the P fertilizer and various soil constituents.

3.2.5 Superphosphate
Superphosphate is formed due to the reaction of rock phosphates and conc. H_2SO_4 as follows:

$$Ca_3(PO_4)_2 + 2H_2SO_4 + 5H_2O \rightleftharpoons Ca(H_2PO_4)_2 + H_2O + CaSO_4 \cdot 2H_2O$$

Mono Calcium
Phosphate

When superphosphate is applied to moist or dry soil before rainfall or irrigation, the monocalcium phosphate gets dissolved in the soil moisture. The roots of the growing plants easily take up this form of phosphorus. However, within a very short time, the solution of monocalcium phosphate is precipitated in the soil pores. Depending on the soil reaction, different soil–fertilizer reaction products are formed that are not soluble in water and become unavailable to plants. Thus, superphosphate is not leached from the soil by rainfall.

When superphosphate is applied to strong acidic soils, it combines with iron (Fe) and aluminum (Al) to form their respective insoluble phosphates and also dicalcium phosphate as represented below:

$$2Ca(H_2PO_4) + Fe_2CO_3 = FePO_4 + 2CaHPO_4 + 3H_2O$$

$\qquad\qquad\qquad\qquad$ (Insoluble)$\quad$ (Dicalcium
$\qquad\qquad\qquad\qquad\qquad\qquad\qquad$ phosphate)

$$Al^{3+} + H_2PO_4^- + 2H_2O \rightleftharpoons 2H^+ + Al \begin{matrix} OH \\ OH \\ H_2PO_4 \end{matrix}$$

$\qquad\qquad\qquad\qquad\qquad\qquad\qquad\qquad$ (Insoluble)

In this way, phosphate becomes unavailable to plants.

When superphosphate is applied to moderately acidic soils, phosphate fixation takes place by the silicate minerals present in soil and becomes unavailable to plants. Al and Fe are removed from the edges of the silicate crystals forming hydroxy phosphates.

$$[Al] + H_2PO_2^- + 2H_2O \rightleftharpoons 2H^+ + Al(OH)_2H_2PO_4$$

In silicate $\qquad\qquad\qquad\qquad\qquad\qquad\qquad$ (Insoluble)
crystal

A part of the reacted phosphate with Fe and Al compounds is also released by anion exchange reactions with OH^- ions.

$$Al(OH_2)H_4PO_4 + OH^- \rightleftharpoons Al(OH)_3 + H_2PO_4^-$$

(Insoluble) $\qquad\qquad\qquad\qquad\qquad\qquad$ (soluble)

Again, when superphosphate is applied to neutral and calcareous soils, the following reaction will take place and P becomes unavailable to plants.

$$Ca(H_2PO_4)_2 + 2Ca^{2+} \rightleftharpoons Ca_3(PO_4)_2 + 4H^+$$

Mono calcium $\qquad\qquad$ (Tricalcium phosphate.
phosphate $\qquad\qquad\qquad$ Insoluble)

$$Ca(H_2PO_4)_2 + 2CaCO_3 \rightleftharpoons Ca_3(PO_4)_2 + 2CO_2\uparrow + 2H_2O$$

$\qquad\qquad\qquad\qquad\qquad\qquad\qquad$ (insoluble)

The insoluble tricalcium phosphate formed is converted in the soil to more insoluble compounds like hydroxyl apatite $[Ca_{10}(PO_4)_5(OH)_4]$; carbonatoapatite $[Ca_{10}(PO_4)_6(CO_3)_2]$, and fluorapatite $[Ca_{10}(PO_4)_6]$. F_2 and ultimately P become unavailable to plants. When superphosphate is applied to alkaline soils (pH > 8.5 or 9.0, Na^+ is the dominant cation), it reacts with Na^+ and forms monosodium phosphate which is highly soluble and becomes available to plants.

$$H_2PO_4 + Na^+ \longrightarrow NaH_2PO_4 \text{ (Soluble)}$$

3.2.6 Rock phosphate

Rock phosphate is stable and insoluble in water. The citrate solubility varies from 5% to about 17% of the total P content. Effectiveness of rock phosphate is determined by its chemical reactivity, which in turn depends on the degree of carbonate substitution for phosphate in the apatite structure. Finely ground apatite phosphate rocks are effective only on acidic soil (pH 6 or below). Whereas, calcined iron-aluminum phosphate ores with much higher citrate solubilities (60%–65%) of P can be used successfully on neutral and calcareous soils. In situations where the reactivity of available rock phosphate is inadequate, partially acidulated rock phosphate (treating rock phosphate with H_2SO_4) can be used successfully. A granular material containing a mixture of raw rock phosphate and finely ground elemental sulfur has been developed and designated as "Bio super." It is inoculated with the sulfur-oxidizing bacteria *Thiohacillus thioxidans* to ensure the conversion of sulfur to sulfuric acid (S to H_2SO_4). This acid in turn reacts with the rock phosphate, releases soluble P in the soil, and thereby increases P availability to plants.

3.3 Potassic fertilizers

Potassium fertilizer is commonly added to agricultural land to improve the yield and quality of plants growing in soils that are lacking an adequate supply of this essential nutrient. About 90%–95% of potash are being used in agriculture as fertilizer. K is the third essential plant macronutrient after N and P. It is vital to many important plant physiological processes including photosynthesis (creation of energy), water and nutrient uptake, and overall crop quality. To ensure healthy and nutritious plant growth, an adequate supply of K must be maintained in the soil. Practically all K fertilizers are water-soluble. Different potassic fertilizers essentially consist of K in combination with chloride, sulfate, nitrate, polyphosphates, etc. The most common type of potash is potassium chloride (KCl). KCl is also known as muriate of potash (MOP) or sylvite, a naturally occurring mineral.

The common K fertilizers are completely water-soluble and, in some cases, have a high salt index. Consequently, when placed too close to seed or transplants, they can decrease seed germination and plant survival. Fertilizer injury is most severe on sandy soils, under dry conditions and with high fertilizer rates, especially N and K fertilizers. Some crops such as soybeans, cotton, and peanuts are much more sensitive to fertilizer injury than corn. Placement of the fertilizer in a band approximately 3 inches to the side and 2 inches below the seed is an effective method of preventing fertilizer injury. The supply of potassium to plants is expressed as ionic (K^+) percentage instead of K_2O percentage in potassic fertilizers. The entire requirement of about five million tonnes of potassic fertilizers are met through imports, as India does not have commercially viable sources of potash. The country is totally dependent on import of potassic fertilizers (Indian Fertilizer Industry at a Glance in 2012—13).

3.3.1 Muriate of potash or potassium chloride (KCl)

The word *muriate* is derived from muriatic acid, a common name for hydrochloric acid. Fertilizer grade muriate contains 50%—52% K (60%—63% K_2O) and varies in color from pink/red to white depending on the mining and recovery process used. Muriate of potash is generally marketed in five particle sizes: special standard, white soluble, standard, coarse, and granular, of which coarse and granular size fractions of muriate of potash are widely used. It is manufactured from potash-bearing minerals namely carnallite—$KCl.MgCl_2.6H_2O$ (17.0% K_2O); sylvite—KCl, sylvinite—KCl and NaCl mixture, etc. Recovery of KCl from sylvinite ore is made by the mineral floatation process or by solution of KCl, followed by recrystallization. Floatation process is based on the differences in the specific gravities of KCl and NaCl. With KCl having lesser specific gravity it floats atop of NaCl, and is thereby separated out. Whereas in the recrystallization process, the difference in temperature-solubility relationships of chloride salts of K and Na is the fundamental principle of this method. The solubility of KCl increases rapidly with a rise in temperature, whereas NaCl solubility varies only slightly over a wide temperature range.

3.3.2 Sulfate of potash (K₂SO₄)

K_2SO_4 is rarely found in a pure form in nature. Instead, it is naturally mixed with salts containing magnesium, sodium, and chloride. These minerals require additional processing to separate their components. Historically, K_2SO_4 was made by reacting KCl with sulfuric acid. Potassium sulfate is a white salt which contains 41.5%—44.2% K (50%—53.2% K_2O). A number of processes like Langbeinite, Trona, Mannheim, Glaserite processes, etc., produce it.

3.3.3 Langbeinite process

Production from Langbeinite minerals ($K_2SO_4.2MgSO_4$, 22.6% K_2O) occurs according to the following reaction:

$$K_2SO_4.2MgSO_4 + 4KCl \longrightarrow 3K_2SO_4 + 2MgCl_2$$

3.3.4 Trona process

Burkite ($Na_2CO_3.2Na_2SO_4$) reacts with potassium chloride to form glaserite ($Na_2SO_4.3K_2SO_4$) and then with KCI brine to give potassium sulfate.

3.3.5 Glaserite process

The following reaction takes place in the production of sulfate of potash from sodium sulfate (Na_2SO_4) and KCI with glaserite as an intermediate product.

$$4Na_2SO_4 + 6KCl \longrightarrow Na_2SO_4.3K_2SO_4 + 6NaCl$$
$$(\text{Glaserite})$$

$$Na_2SO_4.3K_2SO_4 + 2KCl \longrightarrow 4K_2SO_4 + 2NaCl$$
$$(\text{Glaserite})$$

3.3.6 Reactions in soil

Both chloride and sulfate of K are soluble in water; on application to the soil they ionize into K^+, Cl^-, and SO_4^{2-} ions. The released K^+ ions from the fertilizer get adsorbed on the soil colloids and are available to the plant through cation exchange reactions. Based on the chemistry of chloride in the soils, it is concluded that under acidic soil conditions, Cl^- ion replaces the OH^- ions associated with the free iron oxides and therefore, in such soils, muriate of potash is likely to give a greater response than K_2SO_4. Besides, the Cl^- ions are less strongly retained on soil colloids than the SO_4^{2-} ions. In alkaline soils, when KCI is applied, the accumulation of Cl^- ions creates toxicity to plants. Therefore, in potash deficient soils with alkaline reaction, it should be applied along with organic matter. The application of potassic fertilizers has little or no effect on soil reaction.

3.4 Toxicology of chemical fertilizers

Nowadays, enhanced fertilization to enrich soil fertility and crop productivity without knowing its deleterious effects has often negatively affected the biogeochemical cycles. This has polluted almost every part of our ecosystem due to their long-term persistence in the soil and water [10,52]. The harmful effects of these chemical fertilizers starts from the manufacturing process. The products and byproducts of chemical fertilizers are some toxic chemicals or gases like NH_4, CO_2, CH_4, etc., which cause air pollution [10,51]. The untreated industrial wastes are disposed into nearby water bodies, which triggers the devastating effect of chemical accumulation in water bodies, leading to water eutrophication. These toxic agrochemicals also percolate into the soil and degrades the soil health and quality, thereby causing nutrient imbalance, and hence soil pollution [10,50]. These agropollutants also have impacts on human health because of their long residual efficacy and accumulation through food chain, which is figured as being responsible for dreadful human health injuries [53]. Therefore, it is high time to realize that this crop production input depletes our environment and ecosystem. Hence, its continuous use without taking any remedial measures to reduce its use will one day deplete all the natural resources, and as well threaten the entire life forms on Earth. The adverse effects of these synthetic chemicals on human health and environment may be reduced or eliminated by adopting new agricultural methodologies which involves shifting from chemical intensive agriculture and embracing the use of organic inputs like manure, biofertilizers, biopesticides, slow-release fertilizers, nanofertilizers, etc. [10]. All these organic farming practices help to improve the application efficiency as well as usage efficiency of the chemical fertilizers by the crops in an ecofriendly manner by offering a healthy natural ecosystem for the present as well as future generations.

4. Impact of chemical fertilizers on environment

Fertilizer usage is no doubt beneficial to plants in supplementing deficient nutrients, but it should be used in optimum levels required for the soil to produce nutrition-rich and chemical-free agricultural products without deteriorating the quality of our natural resources [28]. There is no denying the fact that enhanced agricultural productivity we achieved in the past century had largely been accomplished through the adequate supply of fertilizers [18]. Such intensive use of chemical fertilizers rather than the recommended dosage by the world agricultural systems to enhance productivity resulted in crucial environment pollution (soil, water, air pollution) leading to decreased food quality, soil degradation, loss of soil porosity, micronutrient deficiency in soil, eutrophication, and has affected biogeochemical cycles and caused toxicity to beneficial microbial flora present in the soil [10,50,53]. However, currently we have reached a stage where all these synthetic chemicals have their drastic side effects in their long run; hence, cited below are certain vital facts defining the negative effects of fertilizers on the environment and humans.

4.1 Effects of chemical fertilizers on soil pollution

The soil is a highly complex natural layer of dynamic ecosystem that harbors micro and macro flora bearing tremendous range of niches and habitats comprising most of the Earth's genetic diversity [4,48]. It consists of both inorganic nutrients and organic mixtures of soil, constituting minerals, gases, liquids, and organisms that together support and maintain life. The soil is the basis of agriculture, which is responsible for many vital functions like energy flow, nutrient recycling, mediator in gas exchange, transfer of nutrients and water, detoxification of pollutants, regulator of water quality, habitat for soil organisms, etc. [39]. The proportions of silt, sand, and clay determine the soil texture, and it also has a number of implications because it affects the water holding ability of soil. The organic matter present in the soil acts as a buffer against the forces of compaction; hence, it reduces the hard setting via improving water infiltration through the soil [21,48]. Continuous cropping and overuse of chemical fertilizers can diminish organic matter content of the soil very rapidly, leading to soil structural decline and soil acidification, thereby, affecting beneficial organisms, stunting plant growth, causing variations in the soil pH, pest proliferation, and even contributing to the undesirable release of greenhouse gases (Fig. 12.2) [10,48].

Likewise, long-term persistence of fertilizers in the soil strongly alters activities of soil enzymes and soil functions like rhizode position, organic carbon and moisture content [48]. Thus, soil health management plays a pivotal role for ensuring sustainable agricultural productions and maintenance of biodiversity. The negative impacts of agro pollutants on soil health are as follows.

4.1.1 Soil acidification

The soil structure is regarded as a prime indicator in agricultural productivity. Soil acidification is a common matter of concern in long-term crop production [48]. It is simply the building up of hydrogen cations, which reduces the soil pH and occurs chemically when a proton donor is supplemented to the soil. The donor can be an acid like nitric acid, sulfuric

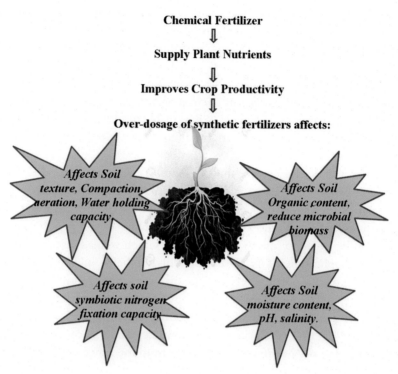

Figure 12.2 Effects of chemical fertilizers on physical, chemical, and biological properties of soil.

acid, or carbonic acid [15]. Soil acidification is a natural phenomenon in regions with medium to high rainfall where leaching gradually acidifies soil over time. Intensive agriculture enhances soil acidification through rapid and unconscious fertilization without proper soil testing [50]. The major fertilizer element N is the main nutrient affecting soil pH. Soil can become more acidic or alkaline depending on the nature and type of N fertilizer used [10,50,51]. Alteration in the soil pH may be advantageous or detrimental depending on the starting pH of the soil and the range and rapidity of the pH change. Macronutrients like N, P, K, and S have major effects on pH as they are added in much larger quantities to soil than micronutrients.

4.1.1.1 Role of N, P, and S in soil acidification

N is a vital nutrient because it is a major component of the genetic material, pigment molecule—chlorophyll, energy-transfer compound—adenosine triphosphate, and amino acids, which comprise the basic structural and functional units of plant cells. Soil nitrogen exists in three forms: (i) organic

N compounds, (ii) ammonium (NH_4^+) ions, (iii) nitrate (NO_3^-) ions [10]. About 90%—95% of the potentially available N in the soil is in organic form as plant or animal debris and also present in living soil microorganisms like bacteria. However, this N is not in a readily available form to plants, but some can be converted to available forms by soil microbes. Majority of plant available N sources are in their inorganic forms NH_4^+ and NO_3^-, whereas a trace amount of organic N is available to plants in soluble organic compounds like urea. Ammonium ions bind to the soil's negatively charged cation exchange complex (CEC), whereas the nitrate ions carry negative charges, and so do not bind to the soil solids. These nitrate ions dissolve in soil water or are precipitated as soluble salts under dry conditions.

In agricultural systems, the form and fate of N in the soil–plant system is probably considered as the key factor of variability in soil pH. N supplements can be added to soils in various forms, which we had discussed above [66]. Overapplication of N to the agricultural fields in large quantities destroys the natural balance between the other three macronutrients, which in turn affects the micronutrients, thereby, resulting in depletion of crop yields. The key molecules of N supplements which alter soil pH are the uncharged urea molecule, anion nitrate, and cation ammonium. The conversion of N from one of these forms to the other also involves the generation or consumption of acidity [7].

P is a very stable, least mobile, and minimum available nutrient for plants in soil; hence, it binds quickly with soil minerals. The possibility of P leaching out in the soil is extremely low, except under high rainfall conditions or during soil erosion. P fertilizers contribute to soil acidity principally through the release or gain of H^+ ions by the phosphate molecule depending on soil pH [7,66]. Sulfur (S) is also an essential nutrient required by plants for optimum production. Soil organic matter is the primary source of plant S; hence, sandy soils with low humus content are more prone to S deficiency. Plants mostly take up S in the sulfate form but when compared to N, the uptake of S by plants is very low. S fertilizers, added to soil, affect soil acidity predominantly through the release of H^+ ions by the addition of elemental S or thiosulfate. K is not required by plants in large quantities and is rapidly taken up by plants. Hence, it contributes little to environmental threat and has no effect on soil acidification [7,10,66].

4.1.2 Heavy metal accumulation in the soils

Repeated application of fertilizers blindly may result in the accumulation of toxic heavy metals such as arsenic (As), cadmium (Cd), and uranium (U) in the soil and plants [10,57]. These metallic elements are considered systemic toxicants that are known to induce life threatening health problems even at lower levels of exposure. Even though N and P fertilizers do not contain substantial levels of Cd but its application in agricultural lands can trigger increased levels of Cd, As, chromium (Cr), and lead (Pb) in soil and dramatically decreased soil pH that causes desorption of heavy metals from the soil matrix [3,10]. Manures, atmospheric deposition, and sludges are the supreme sources of Pb in the agricultural soils. Chemical fertilizers like lime and triple superphosphate contain not only major elements necessary for plant nutrition but also trace elements like Cd and As [3,10,57]. These toxic metals not only pollute the soil but also get accumulated in fruits, vegetables, and grains via the food chain, and eventually reach humans and animals, causing serious health problems. The animals fed with crops grown in the soils with elevated heavy metal concentrations may also have accumulation of high levels of toxic metals in the beef and poultry [36]. Nowadays, the frequent misuse of fertilizers in agriculture causes severe environmental and public health deterioration, thus causing a great concern internationally.

4.1.3 Emission of greenhouse gases

Fossil fuels and related uses of coal and petroleum are the primary sources of greenhouse gases (GHGs). Agriculture is the second most important contributor of GHGs which are generated through feed production, intensive food production with the aid of chemical fertilizers, flooded paddy rice production, and deforestation when farmers are too ambitious to increase cultivated areas [15]. In the agriculture sector, the production of synthetic fertilizers releases far more GHGs than expected and its contribution mainly comes from the production of nitrogenous fertilizers leading to global climatic impacts. Soils have the capacity to generate or store GHGs, depending on how they are treated or managed with synthetic fertilizers [46]. The major GHG emissions associated with the production of N fertilizer are (i) CO_2 emissions when the natural gas is combusted as part of ammonia synthesis and (ii) N_2O emission during nitric acid production. The GHGs promote global warming, deplete the protective ozone layer, cause significant climate change, rise in sea levels, melting the ice, increasing ocean acidification, extreme weather events, and other severe natural and societal global implications on the entire ecosystem [46]. For P fertilizers, CO_2 is emitted during the combustion of fossil fuels and transport of raw materials during the various production processes. In the case of urea fertilizer, only minimal amount of CO_2 emissions occur during the process of urea hydrolysis in the soil after the application of fertilizers.

During the step-by-step production processes of inorganic fertilizers, large amount of GHGs are expelled in three different ways as explained below [46]:

(i) The major CO_2 emissions are from the energy-intensive production process

(ii) CO_2 emissions from the energy used in the transport of raw materials and fertilizers from the production facility to farmlands

(iii) It is one of the common modes of GHG emissions seen in large-scale farming in industrialized countries where CO_2 is emitted from energy use in machinery required for fertilizer application. This is applicable only when the fertilizer spreading was done with the aid of machinery. The storage and emission capacities of GHGs by the soil are quite large and adversely affect the global climatic change.

4.1.4 Accumulation of mineral salts of fertilizers in soils

The overuse of chemical fertilizers in farming causes accumulation and concentration of mineral salts of fertilizers in soil leading to long-term soil compaction and degradation [10,42]. The excess mineral salts alter the physical properties of the soils such as bulk density, reduction in soil volume, penetration resistance, and soil compaction which result in decreased crop yield and increase in production cost [26,42]. The fine and structureless salts present in the fertilizers cover the soil surface and hinder water percolation. As a result, a hard and impermeable layer develops. The resulting soil compaction causes severe complications like extreme soil strength, poor aeration and drainage, limited root growth, which affect the natural nutrients/water uptake capability of plants and reduce soil organic matter [42]. The accumulation of salts indirectly contribute to crop tip browning, lower leaf yellowing, wilting, and crop lodging due to the inability of water absorption [10]. The excessive usage of chemical fertilizers consequently disrupts the biotic activities of the soil by restricting or hindering the free aeration and water percolation in the soil and ultimately resulting in stunted plant growth and low productivity.

4.1.5 Effects on soil microbial community

Fertilization influences both the above and the below-ground microbial biomass. Soil microbes like bacteria, fungi, and archaea play a crucial role in catalyzing the transition between compounds of N, P, and C, thereby benefiting the plant uptake and microbial population [48]. Bacteria belong to the most abundant microbial community in the soil followed by acti-nomycetes, fungi, algae, and protozoa. They play essential roles in fertilizing the soil by decomposing organic matter and via nutrient cycling. Soil microflora play a prime role in ecosystem functioning and regulation of key processes like carbon and nitrogen cycles [70]. The soil microbes lock carbon into the soil for long periods, which enhances soil fertility and water-retaining capacity; hence, microbes are defined as the producers of soil organic carbon [33]. The microbial biomass in the soil mainly consists of bacteria and fungi which constitute nearly 5% of total organic matter of the soil. Hence, the input of inorganic fertilizers into the soil reduces the fungal/bacterial biomass ratios whereas a reverse effect can be seen in the case of organic fertilization [48]. The soil microorganisms are highly sen-sitive to various soil properties like nutrient content, pH, moisture content, organic N, C content, etc., that occur due to intensive fertilization. Bacteria and actinomycetes can somehow withstand soil disturbances due to excess fertilization than the fungal populations, which cannot survive easily in such extreme conditions [48,59]. It is well known that soil microbial commu-nities can be a potential ecological indicator of soil quality. The alterations in soil pH induced by the application of N alone, N plus P, and other chemical fertilizers is a decisive factor in determining bacterial abundance and abundance of other microbial communities [33,48]. Chemical fertilizers can have short or long-term effects on the soil microflora brought about directly by their action on the organisms and indirectly due to undesirable changes in the soil environment [71]. The consistent changes in the abundant structure, classes, and genera of the bacterial communities due to fertilization are considered detrimental and indirectly affect the biodiversity. For the global conservation and maintenance of biodiversity, the mainte-nance of soil microbial flora needs much attention.

4.1.6 Inhibits symbiotic nitrogen fixation by rhizobia

In early days, farmers totally depended on leguminous crops for fixing atmospheric nitrogen via symbiosis with nitrogen–fixing rhizobia bacteria, in rotation with nonleguminous crops rather than current agricultural strategy of depending on nitrogenous fertilizers [20]. Nitrogen fixation is the conversion of stable atmospheric nitrogen gas into a biologically useful form by N fixing microorganisms by reducing dinitrogen into ammonia with the aid of an enzyme complex, nitrogenase [65]. Leguminous plants possess a highly plastic root system to form nitrogen fixing nodules by means of symbiotic interaction with specialized soil bacteria. Nitrogen-fixing bacteria are considered a subset of plant growth—promoting rhizo-bacteria. Chemical fertilizers hinder the recruitment of rhizobia bacteria to host plant root nodules, hence, lowering the rates of nitrogenase activity which then results in a reduction in overall plant yield at the time of harvest [44]. The biological processes like nodulation and nitrogen fixation by legumes are adversely affected by higher doses of synthetic nitrogenous fertilizers. The N fertilizer, especially nitrate, affects nodulation process mainly by destroying indole acetic acid and decreasing root hair formation plus root hair curling, thereby limiting the attachment of rhizobia on root hairs [44]. The inhibitory pathway of nitrate on nodulation of legumes occurs in four steps: (i) infection, (ii) hinders the development of nodules, (iii) obstructs nitrogen fixation, (iv) affects the integrity of bacteroids [20]. In legumes, high nitrate and ammonium in the range of 3 mM concentration inhibit nodulation whereas lower concentrations of 0.5—2 mM can stimulate nodulation. Excess N and ammonia inhibits initial cell division in the root cortex and adversely affects the integrity of bacteroids in the nodules, which in turn affects nitrogenase, thereby inhibiting the process of nitrogen fixation [44]. Apart from inorganic N, free amino acids, particularly glutamine, asparagine, and ureides produced by nodules may also regulate nodulation by reducing nitrogenase activity. Low N levels promote the formation of lateral roots and nodules, which are susceptible to rhizobial infection and act as the residing sites of N fixing microbes. However, due to the high availability of N, plants promotes lateral root elongation to exploit the available N-source [44,65]. In this case, the root is less susceptible to nodulation and infection through the reduction of the number of nodules that develop thereby, reducing N fixing capacity. The tolerance or sensitivity of symbiosis and host-microsymbiont interaction in legumes are strongly influenced by the level of inorganic fertilizer content in the soil. Apart from environmental consequences, the increased dependence on synthetic nitrogenous fertilizers reduces soil fertility, affects the natural symbiotic N fixing process, and results in unsustainable long-term crop yields.

4.1.7 Reduces the nutritional values of crops

As discussed earlier, overfertilization deteriorates the soil quality and properties due to excess supply of micro/macronutrients which later results in the persistence and accumulation of toxic pollutants in the soil. Over-fertilization decreases both the protein and carbohydrate content of crops thereby depleting its nutritional values and edible qualities [41]. The fruits and vegetables grown on soils overenriched with chemical fertilizers are more prone to pest attacks and infectious diseases [34]. The presence of excess K in soil also decreases the antioxidant compounds, carotene, and vitamin C content in vegetables [63].

4.1.8 The antibiotic resistance genes in fertilizers increase soil antibiotic resistome

The presence of antibiotic resistance genes (ARGs) and antibiotic resistance bacteria (ARBs) in synthetic fertilizers is an unignorable serious issue causing higher risks to human health due to the transfer of ARGs and ARBs from fertilizer to soil which then enter the food chain via crops [61,67]. Long-term application of poultry manure significantly enriches β-lactam and tetracycline-related ARGs in soil. In addition to that, chemical pollutants and heavy metals in fertilizers also favor horizontal transfer of ARGs, thereby causing great risk to humans [12]. The massive and long-term application of both organic and inorganic fertilizers increases soil antibiotic resistome, but adversely affects the natural indigenous ARBs and ARGs of the soil [61].

4.2 Effects of chemical fertilizers on water pollution

The application of synthetic fertilizers to the soil in large quantities coupled with the unfavorable environmental conditions increase the risk of leaching, thereby causing pollution. The pollutants ends up in rivulets, lakes, ponds, rivers, and other water sources through drainage, leaching, and flow, resulting in severe water pollution [10,15]. As a result, relatively high fractions of the applied N may potentially be leached or removed from the root zone into the surface and groundwater. When N fertilizers are applied in ideal climatic conditions, plants use only up to 50% of the applied fertilizer, 2%−20% gets volatilized, 15%−25% reacts with organic compounds in the clay soil, and the remaining 2%−10% gets into surface and groundwater [10]. The basic component of N fertilizer, nitrate, plays a central role in water pollution. It can be found in the form of nitrite (NO_2), nitrogen (N_2), nitrogen oxide (N_2O), and organic N. All these synthetic nutrients enrich the dissolved N content of water bodies, and thus result in adverse effects.

4.2.1 Stimulates eutrophication in water bodies

The most deleterious consequence of the intensive use of fertilizers is water eutrophication. Even the fertilizer nutrients left unused in the soil can run

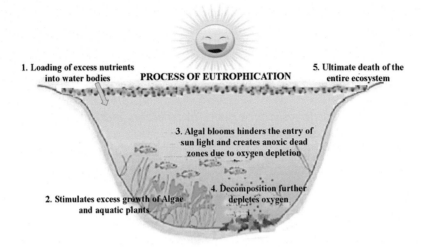

Figure 12.3 The schematic representation of different stages of eutrophication process.

off into water bodies resulting in eutrophication. Overuse of fertilizers and other human activities accelerate the rate and extent of human caused eutrophication through both point-source discharges and nonpoint loadings of limiting nutrients (N and P) into aquatic ecosystems causing detrimental environmental consequences [10,50,51]. Eutrophication is a condition of excessive algae blooms in the water bodies commonly induced by the oversupply of enriched minerals and nutrients (Fig. 12.3). It is often induced by the discharge of nitrate/phosphate fertilizers or sewage into aquatic bodies [13]. Eutrophication indirectly affects biodiversity by altering the species composition and dominance. The effective management of cultural eutrophication is a multifaceted issue which will require the collective efforts of scientists and citizens to reduce synthetic nutrient inputs in order to restore our precious aquatic communities.

   The Impacts of Eutrophication [https://www.pmfias.com/water-pollution-biological-oxygen-demand/] **involves:**

1. Most conspicuous effect of cultural eutrophication is the excess growth of dense blooms of noxious, foul-smelling phytoplanktons that decrease water clarity and quality [13].

2. The algal blooms reduce the growth of other plants in littoral zones and cause dieoffs by limiting light penetration. Furthermore, high rates of photosynthesis associated with eutrophication can also deplete dissolved inorganic carbon and raise water pH to extreme levels [13].

3. The elevated pH due to eutrophication indirectly affects the chemo-sensory abilities of other aquatic organisms resulting in the replacement of desirable fishes by less desirable species.

4. Eutrophication can lead to the killing of aquatic life and the proliferation of unwanted species because the overgrowth of aquatic plants and algae in the water body by covering the entire surface area leads to oxygen depletion, hence, an anoxic (oxygen-free) environment is thus created (Fig. 12.2). It eventually results in the loss of other aquatic living species like fishes, thereby affecting biodiversity [13].

5. When such algal blooms eventually die, microbial decomposition of their biomass results in oxygen consumption, thereby creating the state of hypoxia or anoxic "dead zone" lacking adequate oxygen to support most organisms (Fig. 12.2) [13]. Such hypoxic conditions threaten profitable commercial and recreational fisheries worldwide. The dead zones are found in many freshwater lakes like Laurentian Great Lakes during the summer and are also common in marine coastal environments.

6. Some algal blooms produce noxious toxins like microcystin and ana-toxin-*a* and are indirectly linked with the degradation of water quality, destruction of economically important fisheries, and public health risks [14]. After the consumption of the toxins by small fish and shell-fish, these toxins move into the food chain and can cause adverse impact on aquatic and human lives. The toxigenic cyanobacteria like *Anabaena*, *Cylindrospermopsis*, *Microcystis*, and *Oscillatoria* also tend to flourish dominantly in nutrient-rich freshwater systems and can negatively affect aquatic life by blocking out sunlight and clogging fish gills [13]. Apart from that, cyanobacteria also produces off-flavor compounds such as methylisoborneol and geosmin. They mainly contaminate municipal drinking water systems as well as aquacultures that are used for raising fishes, resulting in large financial losses for state and regional economies [17].

7. Infilling and clogging of irrigation canals with aquatic weeds like water hyacinth due to the overflow of enriched minerals and nutrients is a problematic condition. Management of this massive water conveyance system is highly expensive and it may even pollute the water with decomposed debris and toxins liberated by blooms [13].

8. Poisoning of domestic animals, wildlife, and even humans by toxic algal blooms and cyanobacteria have been documented throughout the world; these toxins enter the food chain through grazing, causing significant public health risk [13].

9. As a results of algal blooms, large quantities of solid materials accumulate in water bodies toward the end of their life cycle, resulting in the reduction of self-purification capacity of water bodies. Consequently, the immensely loaded sediments start to fill the basin and increase the interactions between water and sediment and thereby, facilitate the resuspension of nutrients and could in fact lead to further deterioration of water quality [13].

10. Eutrophication causes prohibition of touristic use of the water bodies and even loss of recreation due to foul odor on the shores as a result of massive algal blooms. The water bodies lose the hygienic and attractive appearance and may even cause skin irritations as well [13].

4.2.2 Enhances ground water pollution

The leaching of vital nutrients from agricultural soil into ground and surface water is a reason for major environmental and public health concern. The field and garden synthetic chemicals, like fertilizers, enter the groundwater in two ways: (i) the chemicals along with rainwater into a stream as runoff, (ii) chemical contamination through leaching, which is the downward movement of a substance through the soil [10,51]. This is particularly problematic in urban locations where hard-surfaced roads permit rainwater to move over them without benefiting the soil. The water in streams replenishes groundwater, so the chemicals are absorbed into the groundwater as well. Fertilizer may also dissolve into the surface water, which in turn recharges the groundwater too.

4.2.2.1 Role of nitrate and heavy metals in ground water pollution

Nitrate is highly soluble and readily available common form of dissolved N present in groundwater. High nitrate accumulation in drinking water, i.e., when nitrate concentration exceeds 50 mg NO_3^-/L, can lead to serious health hazards like blue baby syndrome—acquired methemoglobinemia in infants [50,51]. Heavy metals that are usually present in minimal amount in groundwater around industrial areas are mainly due to excessive application of agrochemicals. The toxic pollutants are retained in the unsaturated zone and eventually contaminate groundwater through irrigation return flow. These biotoxic heavy metals accumulate in crops and subsequently enter into the food chain causing potential human health risks [64].

4.2.2.2 Role of ammonia in ground water pollution

Ammonia in water is an indicator of possible bacterial, sewage, animal waste, and synthetic chemical pollution. Its natural level in ground water and surface water are usually below 0.2 mg/L. Toxic concentrations of ammonia in humans may cause loss of equilibrium, convulsions, coma, and even death. A study also reported that NO_3 content of 50% of well waters were higher than 45 mg/L, which was noted as critical value for NO_3 pollution in waters [10]; https://water-research.net/index.php/ammonia-in-groundwater-runoff-and-streams). Elevated ammonia levels in water bodies affect the hatching and growth rates of fishes and harm aquatic life adversely. Ammonia toxicity is one of the foremost reasons of unexplained losses in fish hatcheries. Excess ammonia may accumulate in the fishes leading to alterations of metabolism, increase in body pH, induced changes in gill tissues resulting in hyperplasia (the condition of swelling and clumping of gill filaments), thereby leading to decrease in breathing ability of fishes. Industrial development and new technologies of fertilization have increased the depth of groundwater pollution; hence, water disinfection process like oxidation and chlorination of groundwater is mandatory. Bromate is an anionic contaminant byproduct generated during such disinfection process of groundwater for drinking purpose (https://www.who.int/water_sanitation_health/dwq/chemicals/ammoniasum.pdf; [10]). The current agronomic practices consume large amount of fertilizers and irrigation water through drip irrigation. In areas with water scarcity, the bromate content in soil and water is expected to increase over time. The maximum feasible amount of nitrate that does not impede groundwater quality critically depends on unconscious agricultural practices and other external factors.

4.2.3 Enhances surface water pollution

P tends to attach to soil particles and hence moves freely into surface-water bodies by runoff. As earlier discussed, the overload of macronutrients (N and P) harms free water by stimulating the growth of algal blooms and other aquatic plants through eutrophication [10]. Since surface water often discharges into ground water by percolation, there is also a concern about high concentrations of P in ground water, affecting water quality [50].

4.2.4 Enhances pollution of marine ecosystem

Fertilizers are commonly used to boost plant growth but these substances are often toxic and can create undesirable organic reactions damaging marine ecosystems. When these nutrients are used to fertilize farmlands, they eventually make their way into rivers, streams, and oceans due to leaching or runoff after heavy rainfall. The marine ecosystems are usually rich in minerals. However, when such high levels of synthetic nutrients are released into it, some algae species explode in growth [10,37]. Eutrophication can also be a reason for explosive blooming of algae. These growth explosions, depending on the type of algae affected, are known as red or brown tides, which cover the surface of the sea and are responsible for huge losses in commercial fisheries. As algae blooms massively, it excretes harmful toxins in large amounts and can easily poison the marine life [37,45]. Due to great deal of oxygen demand by algae blooms, oxygen depletion of water occurs, thereby, transforming the whole area into dead spots where normal marine life can no longer flourish or live at all. When such algal blooms eventually die the microbial decomposition of their biomass again results in oxygen consumption, thus creating a hypoxic condition [13]. This mineralization depletes the oxygen content in the water and there is no other substitute to compensate the loss; hence, solid materials accumulate in water bodies toward the end of algal life cycles [37]. It can still take many years for a particular ecosystem to return to its former healthy condition. These long-lasting, damaging effects can also harm neighboring ecosystems that rely on the various fish and crustaceans to survive.

4.2.4.1 Marine eutrophication and fisheries

Another sector affected by marine eutrophication is fishery. The intense algae blooms and mass development of plants and marine organisms can affect fishing process even mechanically. The decrease of macrophytes due to hypoxic conditions in coastal areas has reduced the reproduction of fish because the developmental stages of their eggs require these plants. Fish reproduction also depends on factors like salinity, reproduction area based on sea depth, water temperature, and oxygen content. Fertilization induced eutrophication threatens all these factors of reproduction, resulting in an unpredictable negative impact on fisheries and aquaculture (https://www.leisurepro.com/blog/ocean-news/damaging-effects-fertilizers-marine-eco-systems/).

4.3 Effects of chemical fertilizers on air pollution

4.3.1 Emission of harmful GHGs results in global warming thereby forcing climate change

The uncontrolled application of fertilizers to enhance the rates of crop productivity and quality results in significant air pollution by generating numerous harmful GHGs responsible for the depletion of the protective ozone layer thereby exposing humans to harmful UV radiations [10]. GHGs are radioactively active gases in earth's atmosphere, which radiate energy in all directions and are responsible for greenhouse effect (GHE).

The earth's natural GHE is able to support life forms, but human activities like burning of fossil fuels, deforestation, and overfertilization have accelerated the GHE and initiated global warming [10,50,51]. Agriculture accounts for nearly 60% of anthropogenic N_2O emissions and the GHG emissions like CO_2, CH_4, and N_2O during the manufacture of nitrogenous fertilizers. The excess use of nitrogenous fertilizers triggers the emission of nitrogen oxides like NO, N_2O, and NO_2 that result in severe air pollution [72]. After CO_2 and CH_4, N_2O possesses effective global warming potential of 310 times more than that of CO_2; hence, it is assigned the third most important GHG [52]. Apart from these GHGs, other gases like hydrogen sulfide, methane, water vapor, and chlorofluoro hydrocarbons like halon gases are also emitted during the process of fertilization [10]. The main concern regarding the combinatorial effect of all these GHGs has to do with global warming and the role of nitrous oxides in stratospheric ozone layer depletion that consequently leads to atmospheric "holes," thus, exposing organisms to excessive harmful UV radiations [49]. The human enhanced GHE by GHGs is an excellent cause of global warming which indirectly affects climatic change. Since all systems in the global climate system are interconnected, adding heat energy toward this causes the global climate as a whole to change. Methane (CH_4) is also a potent GHG, its emission from transplanted paddy fields via agriculture is also a serious concern. Fertilization also increases CH_4 concentration when ammonium-based fertilizers are applied in the fields [10,49]. All these emissions contribute to global industrialization, which has raised atmospheric CO_2 levels from 280 to 400 ppm in the last 150 years, thereby concluding the tremendous hike in the human-produced GHGs. The increased amount of GHGs in the atmosphere absorbs more thermal radiations, resulting in warming up of the Earth's surface and its lower atmosphere leading to extra warming called *greenhouse effect*.

4.3.2 Acid rain causes mineral depletion in the soil and aquatic bodies

Ammonia volatilized or emitted from fertilized lands called agri-ammonia vapors gets deposited dry in atmosphere, and after the chemical transformations are oxidized to nitric and sulfuric acids, hence, creating acid rain. It occurs widely due to the washout of oxides of S, N, and other constituents in the atmosphere in the form of rain [52]. Nitric acid rain is derived primarily from power plants, car exhausts, and from gases released by fertilizer usage and production. Its dangerous effects can most clearly be seen in both land and aquatic environments. Acid rain percolates through the soil and causes lowering of soil pH, which mobilizes and leaches away aluminum and other positively charged ions, i.e., nutrient cations from the soil and increases the availability of toxic heavy metals in the soil. Such changes in both the physical and chemical characteristics of the soil reduce the soil fertility, which ultimately results in low agricultural productivity. Productivity is lowered because photosynthetic rate, stomatal conductance, and other morphological characteristics of plants have been affected. Soil quality, which plays a crucial role in maintaining the structural diversity of our ecosystem, is totally disturbed by the acid rain. The acidic rain water when leached through the soil flows into streams and lakes resulting in acidification of water bodies, thereby, causing adverse effects on aquatic organisms, mainly fishes, and indirectly affects human health too (https://www.scientificamerican.com/article/acid-rain-caused-by-nitrogen-emissions/). At various pH due to acidification, different species have diverse tolerance ranges; hence, there are observed changes in the biodiversity distribution which could result in species extinction in aquatic bodies. Acid rain also affects the decomposition rate inside the water bodies because fungi and bacteria are not tolerant of acidic conditions. The impact of acid deposition on stone monuments and buildings has been recognized for over a century. Acid precipitation causes chemical deterioration of carbonate stones, limestone, marbles, cement, and concretes too. It promotes chemical weathering and corrosion of exposed materials like steel buildings, railway tracks, and other structures made of iron. Acid rain, the invisible form of pollution, also causes adverse impact on human health by acidification of food and water supplies (https://www.organicconsumers.org/news/nitrogen-fertilizer-industrial-farming-acid-rain-back). Agriculture is increasingly functioning like a well-managed industrial operation creating serious environmental and health problems.

4.4 Effects of chemical fertilizers on human health

In a developing country like India, to meet the demands of agricultural goods adequately in order to feed the increasing population, huge application of chemical fertilizers is mandatory. The hazards caused by chemical fertilizers on human health and environment highly depend on the dosage, type, and method of fertilizer application [53]. The following are some of the deteriorating health hazards caused by massive fertilization:

1. The excessive and indiscriminate use of toxic agrochemicals causes them to get accumulated in the soil and runoff into water sources, leading to and contamination of the human food chain, soil, water, and air. When chemical fertilizers are used in farmlands, they are absorbed by the plants and transmitted directly or indirectly into fruits and vegetables and later

enter into human/livestock bodies through food chain. These chemicals increase the risk of different types of cancer like leukemia, lymphoma, brain tumor, prostate and colon cancer.

2. Nitrate concentrated water due to excessive fertilization may lead to serious health hazards like methemoglobinemia in infants [50,57] and also possess strong carcinogenic effects. It acts as a risk factor in developing gastric or intestinal cancer and other ailments like goiter, birth defects, and cardiovascular diseases.

3. Heavy metal pollutants like Cd, As, Cr, and Pb are severe pulmonary and gastrointestinal irritants with carcinogenic potential inducing chromosomal aberrations, micronuclei formation, and inhibiting DNA repair [32,53]. Heavy metal poisoning of both soil and ground water as a result of synthetic fertilizers may even cause human health hazards like renal failure, osteoporosis, respiratory infections, etc. [53].

4. Pollution of water bodies by toxic metals such as mercury (Hg) and Pb from industrial wastewaters adversely affects humans and other animals. Hg causes Minamata (neurological syndrome) disease in human beings, whereas Pb interferes with a variety of body processes and is toxic to many internal organs and tissues.

5. Ammonium nitrate present in fertilizers cause other health complications such as eye, nose, and skin irritations; nausea; uncontrolled muscle movements; giddiness; and collapse [50,51].

6. Potassium chloride interferes with nerve impulses and signal transductions by interrupting all vital body functions. It can cause gastric pains, cardiac arrests, dizziness, bloody diarrhea, convulsions, headaches, and mental impairments [53].

5. Conclusion

Fertilizers are considered a necessary evil because they enhance agricultural productivity by restoring the soil nutrients. On the other hand, fertilizers are highly toxic to humans and the environment. The current problems with fertilizer application are (i) Overdosage of synthetic fertilizers, (ii) presence of toxic elements whose concentrations are above permissible limit in fertilizers, (iii) frequent applications of fertilizers without proper intervals. Balanced fertilization with proper soil testing is the only solution to grow healthy and quality crops sustainably for healthy human populations. In order to safeguard human life and environment from the deleterious effects of fertilizers, the prime step to take is to nurture our mother Earth through organic farming system. Organic farming is a holistic production system, which is sustainable for both human health and environment. Therefore, to guarantee sustainable agricultural production and to safeguard the environment, pollution hazards must be avoided while integrated use of nutrient supplements in the form of organic manures, biofertilizers, nanofertilizers, and other slow-release fertilizers should be adopted. Moving back to our ancestor's theory of adopting organic agriculture will create a healthy natural ecosystem for the present as well as future generations.

Acknowledgment

Surya acknowledges the Department of Science and Technology, Government of India (DST), for the award of DST-SERB-National post-doctoral fellowship.

References

[1] Deleted in review.
[2] S.,S. Alinajoati, B. Mirshekari, Effect of phosphorus fertilization and seed bio fertilization on harvest index and phoshorus use efficiency of wheat cultivars, J. Food Agric. Environ. 9 (2011) 388–397.
[3] Z. Atafar, A. Mesdaghinia, J. Nouri, M. Homaee, M. Yunesian, M. Ahmadimoghaddam, A.H. Mahvi, Effect of fertilizer application on soil heavy metal concentration, Environ. Monit. Assess. 160 (2010) 83–89.
[4] L. Avidano, E. Gamalero, G.P. Cossa, E. Carraro, Characterization of soil health in an Italian polluted site by using microorganisms as bioindicators, Appl. Soil Ecol. 30 (2005) 21–33.
[5] A.T. Ayoub, Fertilizers and the environment, Nutrient Cycl. Agroecosyst. 55 (1999) 117–121.
[6] Deleted in review.
[7] P. Barak, B.O. Jobe, A. Krueger, L.A. Peterson, D.A. Laird, Effects of long-term soil acidification due to agricultural inputs in Wisconsin, Plant Soil 197 (1998) 61–69.
[8] Deleted in review.
[9] N.C. Brady, R.R. Weil, The Nature and Properties of Soils, fourteenth ed., Prentice Hall, Upper Saddle River, New Jersey, USA, 2008.
[10] Chandini, Randeepkumar, Ravendrakumar, O. Prakash, The impact of chemical fertilizers on our environment and ecosystem, in: Book: Research Trends in Environmental Sciences, second ed., 2019, pp. 69–86. Chapter: 5.
[11] D. Chen, L. Yuan, Y. Liu, J. Ji, H. Hou, Long-term application of manures plus chemical fertilizers sustained high rice yield and improved soil chemical and bacterial properties, Eur. J. Agron. 90 (2017) 34–42.
[12] Q. Chen, X. An, H. Li, J. Su, Y. Ma, Y.G. Zhu, Long-term field application of sewage sludge increases the abundance of antibiotic resistance genes in soil, Environ. Int. 92 (93) (2016b) 1–10.
[13] M.F. Chislock, E. Doster, R.A. Zitomer, A.E. Wilson, Eutrophication: causes, consequences, and controls in aquatic ecosystems, Nat. Educ. Knowl. 4 (4) (2013) 10.
[14] I. Chorus, J. Bartram (Eds.), Toxic Cyanobacteria in Water: A Guide to Their Public Health Consequences, Monitoring, and Management, E & FN Spon, London UK, 1999.

[15] G.W. Cooke, Fertilizing for Maximum Yield, third ed., English Language Book society/Collins, 1982, p. 457.

[16] D. Cordell, J.O. Drangert, S. White, The story of phosphorus: global food security and food for thought, Global Environ. Change 19 (2009) 292−305.

[17] J.R. Crews, J.A. Chappell, Agriculture and Natural Resources U.S. Catfish Industry Outlook, Auburn University, Auburn AL, 2007.

[18] P. Desai, A. Patil, K. Veeresh, An overview of production and consumption of major chemical fertilizers in India, J. Pharmacogn. Phytochem. 6 (6) (2017) 2353−2358.

[19] J. Dhillon, G. Torres, E. Driver, B. Figueiredo, W.R. Raun, World phosphorus use efficiency in cereal crops, Agron. J. 109 (2017) 1670−1677.

[20] R.C. Dogra, S.S. Dudeja, Fertilizer N and nitrogen fixation in legume−Rhizobium symbiosis, Ann. Biol. 9 (2) (1993) 149−164.

[21] J.W. Doran, M. Safley, Defining and assessing soil health and sustainable productivity, in: C. Pankhurst, B.M. Doube, V.V.S.R. Gupta (Eds.), Biological Indicators of Soil Health, CAB International, Wallingford, 1997, pp. 1−28.

[22] FAO, Resource STAT-Fertilizer. Food and Agriculture Organization of the UnitedNations, 2009, 12.03.2009, http://faostat.fao.org/site/575/Desktopefault.aspx?PageID=575#ancor.

[23] Deleted in review.

[24] Deleted in review.

[25] J.P. Hammond, M.R. Broadley, P.J. White, G.J. King, H.C. Bowen, R. Hayden, M.C. Meacham, A. Mead, T. Overs, W.P. Spracklen, D.J. Greenwood, Shoot yield drives phosphorus use efficiency in Brassica oleracea and correlates with root architecture traits, J. Exp. Bot. 60 (2009) 1953−1968.

[26] M.A. Hamza, W.K. Anderson, Soil compaction in cropping systems: a review of the nature, causes, and possible solutions, Soil Tillage Res. 82 (2005) 121−145.

[27] T.N. Hartley, A.J. Macdonald, S.P. McGrath, F.J. Zhao, Historical arsenic contamination of soil due to long-term phosphate fertiliser applications, Environ. Pollut. 180 (2013) 259−264.

[28] J.L. Havlin, S.L. Tisdale, W.L. Nelson, J.D. Beaton, Soil Fertility and Fertilizers, eighth ed., Publisher: Pearson, 2013, p. 528.

[29] Deleted in review.

[30] Deleted in review.

[31] M. Irfan, M.Y. Memon, J.A. Shah, M. Abbas, Application of nitrogen and phosphorus in different ratios to affect paddy yield, nutrient uptake and efficiency relations in rice (Oryza sativa L.), J. Environ. Agric. 1 (2016) 79−86.

[32] L. Jarup, Hazards of heavy metal contamination, Br. Med. Bull. 68 (2003) 167−182.

[33] C. Johns, Living Soils: The Role of Microorganisms in Soil Health, 2017. http://www.futuredirections.org.au/publication/living-soils-role-microorganisms-soil-health/.

[34] J. Karungi, B. Ekbom, S. Kyamanywa, Effects of organic versus conventional fertilizers on insect pests, natural enemies and yield of Phaseolus vulgaris, Agric. Ecosyst. Environ. 115 (2006) 51−55.

[35] M.J. King, R.J. Forzatti, Sulphur based by-products from the non-ferrous metals industry, in: J. Liu, J. Peacey, M. Barati, S. Kashani-Nejad, B. Davis (Eds.), Pyrometallurgy of Nickel and Cobalt 2009. Proceeding of the 48th Conference of Metallurgists of CIM, Sudbury, Ontario, Cananda, METSOC, Montreal Quebec, 2009, pp. 137−149.

[36] R. Kooner, B.V.C. Mahajan, W.S. Dhillon, Heavy metal contamination in vegetables, fruits, soil and water—a critical review, Int. J. Agric. Environ. Biotechnol. 7 (3) (2014) 603−612.

[37] U. Kremser, E. Schnug, Impact of fertilizers on aquatic ecosystems and protection of water bodies from mineral nutrients, Landbauforschung Volkenrode 52 (2002) 81−90.

[38] H. Lambers, W.C. Plaxton, Phosphorus: back to the roots, Annu. Plant. Rev 48 (2015) 3—22.

[39] W.E. Larson, F.J. Pierce, The dynamics of soil quality as a measure of sustainable management, in: Book: Defining Soil Quality for a Sustainable Environment, Special Publication, Soil Science Society America, 1994, pp. 37—51.

[40] A.M. Manschadi, H.P. Kaul, J. Vollmann, J. Eitzinger, W. Wenzel, Developing phosphorus efficient crop varieties- an interdisciplinary research framework, Field Crop. Res. 162 (2014) 87—98.

[41] H.A. Marzouk, H.A. Kassem, Improving fruit quality, nutritional value and yield of Zaghloul dates by the application of organic and/or mineral fertilizers, Sci. Hortic. 127 (2011) 249—254.

[42] J. Massah, B. Azadegan, Effect of chemical fertilizers on soil compaction and degradation, AMA 47 (1) (2016) 44—50.

[43] F.F. Mendes, L.J.M. Guimaraes, J.C. Souza, P.E.O. Guimaraes, J.V. Magalhaes, A.A.F. Garcia, S.N. Parentoni, C.T. Guimaraes, Genetic architecture of phosphorus use efficiency in tropical maize cultivated in a low-P soil, Crop Sci. 54 (2014) 1530—1538.

[44] A. Nadiatul, R. Mohd, M.A. Djordjevic, I. Nijat, Nitrogen modulation of legume root architecture signalling pathways involves phytohormones and small lregulatory molecules, Front. Plant Sci. 4 (2013) 385.

[45] L. Ngatia, J.M. Grace III, M. Daniel, T. Robert, Nitrogen and Phosphorus Eutrophication in Marine Ecosystems, 2019, https://doi.org/10.5772/intechopen.81869.

[46] C. Oertel, J. Matschullat, K. Zurba, F. Zimmermann, S. Erasmi, Greenhouse gas emissions from soils—a review, Geochemistry 76 (2016) 327—352.

[47] G. Peyvast, P. Ramezani Kharazi, S. Tahernia, Z. Nosratierad, J.A. Olfati, Municipal solid waste compost increased yield and decreased nitrate amount of broccoli (*Brassica oleracea* var. Italica), J. Appl. Hortic. 10 (2) (2008) 129—132.

[48] P. Prashar, S. Shah, Impact of fertilizers and pesticides on soil microflora in agriculture, in: E. Lichtfouse (Ed.), In Book: Sustainable Agriculture Reviews, vol. 19, Springer, Cham, 2016, pp. 331—361.

[49] T. Rutting, H. Aronsson, S. Delin, Efficient use of nitrogen in agriculture, Nutrient Cycl. Agroecosyst. 110 (2018) 1—5.

[50] S. Savci, An agricultural pollutant: chemical fertilizer, Int. J. Environ. Sci. Dev. 3 (2012a) 77—79.

[51] S. Savci, Investigation of effect of chemical fertilizers on environment, APCBEE Proced. 1 (1) (2012b) 287—292.

[52] A. Sharma, R. Chetani, A review on the effect of organic and chemical fertilizers on plants, Int. J. Res. Appl. Sci. Eng. Technol. (2017) 677.

[53] N. Sharma, R. Singhvi, Effects of chemical fertilizers and pesticides on human health and environment: a review, Int. J. Agric. Environ. Biotechnol. 10 (6) (2017) 675—679.

[54] Deleted in review.

[55] S. Shimbo, T. Watanabe, Z.W. Zhang, M. Ikeda, Cadmium and lead contents in rice and other cereal products in Japan in 1998—2000, Sci. Total Environ. 281 (2001) 165—175.

[56] H. Singh, A. Verma, M.W. Ansari, A. Shukla, Physiological response of rice (*Oryza sativa* L.) genotypes to elevated nitrogen applied under field conditions, Plant Signal. Behav. 9 (2014) e29015.

[57] I. Sonmez, M. Kaplan, S. Sonmez, An investigation of seasonal changes in nitrate contents of soils and irrigation waters in greenhouses located in Antalya-Demre region, Asian J. Chem. 19 (7) (2007) 5639.

[58] V. Smil, Enriching the Earth: Fritz Haber, Carl Bosch, and the Transformation of World Food Production, The MIT Press, Cambridge, USA, 2001.

[59] A. Sradnick, R. Murugan, M. Oltmanns, J. Raupp, R.G. Joergensen, Changes in functional diversity of the soil microbial community in a heterogeneous sandy soil after long-term fertilization with cattle manure and mineral fertilizer, Appl. Soil Ecol. 63 (2013) 23–28.

[60] W.M. Stewart, D.W. Dibb, A.E. Johnston, T.J. Smyth, The contribution of commercial fertilizer nutrients to food production, Agron. J. 97 (2005) 1–6.

[61] Y. Sun, T. Qiu, M. Gao, M. Shi, H. Zhang, X. Wang, Inorganic and organic fertilizers application enhanced antibiotic resistome in greenhouse soils growing vegetables, Ecotoxicol. Environ. Saf. 179 (2019) 24–30.

[62] D. Tilman, K.G. Cassman, P.A. Matson, R. Naylor, S. Polasky, Agricultural sustainability and intensive production practices, Nature 418 (2002) 671–677.

[63] R.K. Toor, G.P. Savage, A. Heeb, Influence of different types of fertilizers on the major antioxidant components of tomatoes, J. Food Compos. Anal. 19 (2006) 20–27.

[64] E. Vetrimurugan, K. Brindha, L. Elango, O.M. Ndwandwe, Human exposure risk to heavy metals through groundwater used for drinking in an intensively irrigated river delta, Appl. Water Sci. 7 (2017) 3267–3280.

[65] S.C. Wagner, Biological nitrogen fixation, Nat. Educ. Knowl. 3 (10) (2011) 15.

[66] A. Wallace, Soil acidification from use of too much fertilizer, Commun. Soil Sci. Plant Anal. 25 (1994) 87–92.

[67] M. Wang, P. Liu, W. Xiong, Q. Zhou, J. Wangxiao, Z. Zeng, Fate of potential indicator antimicrobial resistance genes (ARGs) and bacterial community diversity in simulated manure-soil microcosms, Ecotoxicol. Environ. Saf. 147 (2018) 817–823.

[68] H. Yuan, D. Liu, Signaling components involved in plant responses to phosphate starvation, J. Integr. Plant Biol. 50 (2008) 49–859.

[69] Deleted in review.

[70] W.H. Zhong, Z.C. Cai, Long-term effects of inorganic fertilizers on microbial biomass an community functional diversity in a paddy soil derived from quaternary red clay, Appl. Ecol. 36 (2007) 84–91.

[71] C.S. Nautiyal, P.S. Chauhan, C.R. Bhatia, Changes in soil physico-chemical properties and microbial function diversity due to 14 years of conversion of grassland to organic agriculture in semi-arid agroecosystem, Soil Tiliage Res. 109 (2010) 55–60.

[72] S. Shoji, J. Delgado, A. Mosier, Y. Miura, Use of controlled release fertilizers and nitrification inhibitors to increase nitrogen use efficiency and to conserve air and water quality, Commun. Soil Sci. Plant Anal. 32 (7-8) (2001) 1051–1070.

CHAPTER 13

Organic fertilizers as a route to controlled release of nutrients

Hitha Shaji, Vinaya Chandran, Linu Mathew
School of Biosciences, Mahatma Gandhi University, Kottayam, Kerala, India

1. Introduction

Organic fertilizers are naturally available mineral sources that contain moderate amount of essential plant nutrients. Organic fertilizers can be natural (manure and slurry) or processed, such as compost, blood meal and humic acid, natural enzyme-digested proteins, fish meal, and feather meal [1]. Organic fertilizers act as slow-release fertilizers, in a sense, they provide nutrients in lower amount over an extensive time period. Nitrogen (N), phosphorous (P), and potassium (K) are the three major macronutrients important for plant growth. Synthetic or quick-release fertilizers contain high amount of soluble nitrogen, which is easily soluble in water. Whereas, majority of organic fertilizers contain balanced amount of raw nitrogen and thus work as slow-release fertilizers. By their nature, organic fertilizers mitigate the risk of eutrophication, ground water contamination, and overfertilization.

Organic fertilizers have the following advantages:
- Improve soil (microbiological, physicochemical, and biochemical) properties and thus influence soil quality.
- Help in replenishing the loss in organic matter in short- and long-term periods and thus maintain soil fertility.
- They enhance the existing soil nutrients, and thereby healthy growth is achieved with minimum nutrient densities.
- Minimize environmental damage without reducing crop yields and achieve sustainable levels of agriculture production.

2. Natural organic fertilizers

2.1 Organic manure

Organic manures are mostly derived from animal excretion (except in the case of green manure). The nutrient concentrations in manures vary widely

Controlled Release Fertilizers for Sustainable Agriculture
ISBN 978-0-12-819555-0
https://doi.org/10.1016/B978-0-12-819555-0.00013-3

with the kind of animal they are from. They are the main sources of organic fertilizer in agriculture. Manures boost the fertility of the soil by adding nutrients such as nitrogen that can be utilized by microorganisms in the soil. Manure also improves soil structure and increases its water holding capacity. Almost all farmers use organic fertilizers or manures and apply them on the field depending on the chemistry of the soil, type of crop, the season, and the farmer's previous experience and observation [2].

2.2 Peanut hulls manure

Peanut shells are great for mulching (a covering, as of straw, compost, or spread on the ground around plants to prevent excessive evaporation or erosion, to enrich the soil) and to inhibit weed growth. They are a wonderful source of N, P, and K. They are mainly used for the manufacturing of livestock feed; and also known to improve productivity in terms of yield and thought to increase soil fertility. The effect of peanut shells compost on the growth of *Viola tricolor* and marigold was investigated during 7 months and the results showed that peanut shells compost had more effects on growth properties like height, stem, and leaf dry weight in comparison to control [3].

2.3 Poultry manure

Poultry litter is one of the best organic fertilizers available and is an extremely valuable resource. It improves soil fertility and enhances the development of the roots system and the vigor of the plants and makes them less susceptible to diseases and pest attacks. Poultry manure mineralizes fast in soil and produces a lot of heat; hence, it is not advisable to use it during warm seasons. The adverse impacts resulting from land application of poultry manure may be prevented by implementation of effective best management practices (BMPs). A study was conducted to detect the efficacy of different levels of poultry manure (PM) on growth and yield of *Citrullus lanatus*. The results showed that application of poultry manure significantly enhanced growth parameters like vigor and number of fruits during the two seasons [4].

2.4 Fish manure

Dried and ground fish is a valuable fertilizer and it has a broad range of applications compared to the other manure types. This improves plant resistance to pests such as bollworms and nematodes, and the quality of the fruits (coloration and rate of dry matter). Fish manure is suitable for crops like lettuce and onion. The major disadvantage of fish manure is that it promotes diseases during rainy season and causes a pungent smell. A study was conducted to establish the chemical composition of fish fecal waste determined from fresh manure samples collected at 12 commercial farms growing rainbow trout *Oncorhynchus mykiss* in Ontario, Canada. The result indicated that fish manure tended to have a greater content of Mn, Cd, Cr, Pb, Fe, and Zn than most other livestock manures, but had lower levels of As, Se, Co, and Ni. The data from this study indicate that fresh fish manure is similar to other livestock manures in its chemical composition, and should be suitable for use as an agricultural fertilizer [5].

2.5 Cattle manure

Cow dung, also known as cow manure, is the excretory waste of bovines; like domestic cattle, buffalo, bison, and yak. Cow dung is the undigested residue of plant matter which has passed through the animal's gut. The resultant fecal matter is rich in minerals and is commonly used in urban agriculture. Composted cow manure fertilizer makes an excellent growing medium for garden plants. They are mixed into the soil or used as top-dressing for plants and vegetables as a nutrient-rich fertilizer. They are commonly used as starting inputs as they mineralize slowly. There is evidence that animal manure can increase the pH of acid soils. Manure-amended soil had significantly higher pH than unamended soil [6].

2.6 Horse dung manure

Horse manure is a good source of nutrients and a popular additive to many home gardens. Horse dung manure is highly valued by farmers because composting of horse manure makes the compost pile become super charged and also increases soil fertility, regeneration, and high quality yields. It makes a suitable and inexpensive fertilizer for plants. Horse manure can give new plants a jump start while providing essential nutrients for continual growth. Composting of horse manure does not require any special tools or structures. In fact, it can easily be composted by mixing with a shovel or pitchfork. It averts the negative effects of salinity and improves the capacity of soil water retention. This manure is recommended for crops like lettuce, tomato, and mint because it increases the crop yield with a lasting presence in the soil [7].

3. Processed organic fertilizers

3.1 Bone meal

The traditional production of mineral nutrients like N and P fertilizers is unsustainable due to its reliance on fossil fuels in the case of N, and on limited mineral resource stocks in the case of P. Thus, the use of alternative or complementary fertilizers that originate from organic waste materials is gaining interest. Bone meal decomposes slowly and releases phosphorus gradually. Studies were made on the use of cod bone meal, as a nutrient source for bioremediation [8]. In contaminated cold region soils microbial biodegradation of petroleum hydrocarbon is often limited due to lack of available nutrients, particularly nitrogen. Addition of nitrogen to periglacial soils shows a significance response to microbial activity, although excess level of nitrogen concentration can inhibit biodegradation by decreasing soil water potential. In the case of water-soluble inorganic fertilizer, they quickly partition into soil water, increasing the salt concentration, and imposing an osmotic potential. Hence the best strategy that can be used to avoid microbial inhibition is the use of controlled-release fertilizers. Nitrogen mineralization from cod bone meal was greater at 20°C at pH 6.5 and 7.5. Animal byproducts (ABP) are rich in nutrients and energy. Meat bone meal (MBM) contains considerable amounts of nutrients and can be used as a potential organic fertilizer for agricultural crops. MBM was compared to conventional mineral NPK fertilizers in an experiment conducted in spring barley (*Hordeum vulgare*). The grain yield of the cereal species supported by MBM did not differ from the yield obtained with the mineral fertilizer of any N level; showing that the MBM and mineral fertilization showed no differences in quality in terms of grain weight, test-weight, protein content, and protein yield [9]. The study assessed and compared the environmental impact of using MBM as fertilizer with that of using chemical fertilizer. The results of the study indicated that the nutrient recovery and chemical fertilizer replacement had lower emissions of greenhouse gases and acidification [10].

3.2 Cottonseed meal

Cottonseed meal is the byproduct of oil extraction from cotton seeds. They are mainly used for acid-loving plants such as rhododendrons, blueberries, and azaleas. Several methods are used to extract cottonseed oil such as mechanical extraction, direct solvent extraction process, prepress solvent extraction resulting in different types of cottonseed meal having difference in protein, fiber, and oil content. Naturally obtained cotton seed meal fertilizers are applied prior to planting to treat high soil pH to replace depleted trace elements in the soil. This fertilizer has an N to K ratio of 6:4. Due to its high nutrient content it can be used as a perfect nitrogen fertilizer. An experiment was conducted to evaluate the potential of different organic wastes from the agri-food industry for growing greenhouse tomato (*Lycopersicon esculentum* Mill. "Vision") transplants. Prior to transplanting the soil was thoroughly mixed with cottonseed fertilizer and other organic materials. Cottonseed fertilizer produced the best growth by significantly increasing the shoot dry weight by 57%—83% compared with nonfertilized plants [11].

3.3 Blood meal

Blood meal is one of the main fertilizers allowed to be used in organic farming. Dried blood from cattle slaughterhouses can be used as organic N fertilizer because it contains about 10%—13% organic N. The main component of the fertilizer is hemoglobin, which is characterized by the presence of a prosthetic group containing iron (Fe). The blood needs to be dried before being used as blood meal and is spread on gardens to deter pest animals such as rabbits. These animals smell the blood and are repelled by the odor. Several drying methods such as, oven drying, drum drying, flash drying or spray drying, and solar drying, are available to make inert powder of blood meal. In an experiment, blood meal was incubated in the soil for a period of 1 year and organic matter in the soil composition was evaluated at regular time intervals using a sophisticated technique like isoelectric focusing (IEF) and humification parameters. The results showed that the availability of the Fe increased during the incubation period due to the progressive degradation of the prosthetic group and the successive chelation of the Fe from the humic substances [12].

3.4 Seaweed

It is a broad spectrum fertilizer that is rich in beneficial trace metals and hormones that stimulate plant growth. Seaweed is low in cellulose, so it breaks down quickly and rich in carbohydrates which are essential building blocks in growing plants. The main benefit of using seaweed fertilizer is that it does not produce any unpleasant odor as that of fish emulsion but is more costly. It is either applied to the soil as mulch or can be added to the compost heap, where it is an excellent activator. A less serious potential problem with seaweed is that it increases concentration of salt content, which can upset the balance of salt in soil. To reduce the salt content, seaweed are hosed down before adding to soil or left in rain water for desalination. A study was conducted to detect the effect of seaweed liquid fertilizers (SLF) of *Sargassum wightii* and *Caulerpa chemnitzia* on growth and biochemical constituents of *Vigna sinensis*. The seeds soaked with aqueous extract of seaweeds exhibited a 100% germination when compared to the water-soaked controls [13]

3.5 Wood ash

Wood ash from bonfire is known to be a potential source of potash and lime for agriculture use for many years. Crop nutrient quality is generally improved by applying ash because it provides many of the trace elements that plants need to thrive. Wood ash fertilizer is best used either lightly scattered or by first being composted along with other organic matter because it will produce lye and salts if it gets wet. This lye and salt in larger amounts can cause burn to plants. Composting of wood ash allows the lye and salt to be leached away. Depending on the type of wood, the nutrient composition of wood ash varies. Ashes made from burning hard woods like oak and maple have high nutrients and minerals compared to soft wood. The salt in the wood ash is useful as pest control to kill bothersome pests like snails, slugs, and some kinds of soft-bodied invertebrates, by simply sprinkling it around the base of the plants.

4. Biofertilizers

Biological fertilizer is a substance which contains living microorganisms which when applied to plants either on the surfaces or soil has the ability to colonize the rhizosphere or the interior of the plant. They promote growth by increasing the supply or availability of primary nutrients to the host plant. Biological fertilizer boosts the nutrient composition of soil through the processes of nitrogen fixation and solubilizing mineral ions and thereby stimulates plant growth through the synthesis of growth-promoting substances. The use of biofertilizer is expected to reduce the use of chemical fertilizer and synthetic pesticides. They accelerate microbial processes in the soil which augment the availability of nutrient in a form easily assimilated by plants which means that the microorganism present in the biofertilizer helps to restore the soil's natural nutrient cycle and build soil organic matter. Use of biofertilizer has become one of the important components of integrated nutrient management, as they are costeffective and also a renewable source of plant nutrients to supplement the chemical fertilizers for sustainable agriculture. Additionally it can provide healthy plant growth, while enriching the sustainability and the fertility of the soil [14].

Benefits of biofertilizer are:

1. These are means of fixing the nutrients available in the soil.
2. Since a biofertilizer technically contains a living organism, it can symbiotically associate with plant roots.
3. Microorganisms can readily and safely convert complex organic material into simple compounds, so that they are easily taken up by the plants.
4. Microorganisms function in long duration, causing improvement of the soil fertility.
5. It maintains the natural habitat of the soil.
6. It increases crop yield by 20%–30%, replacing chemical N and P, thereby stimulating plant growth.
7. It can also provide protection against drought and some soil-borne diseases.

The interest in biofertilizers is increasing day by day due to their potential application in sustainable agriculture. However, compared to the biofertilizer formulations used in ancient times many of the products that are currently available are often of very poor quality and quantity. Creating a biofertilizer is a crucial multistep process that results in one or several strains of microorganisms encapsulated in a suitable carrier.It involves providing a safe environment to protect the microorganisms from the often-harsh environmental conditions during storage and ensuring survival and establishment after their release into the soil. There were also issues in the development and production of a biofertilizer such as controlling the product quality at each stage of processing [15]. Hence, a key constraint in successfully obtaining an effective inoculant is overcoming difficulties in formulating a viable and userfriendly final product and maintaining these microbial cells in a competent state [16].

Some important groups of biofertilizer include: microbial fertilizer containing living microorganisms which when applied to the plant, seed, or soil colonize the region of the soil or the interior of the plant and directly influence root secretions and associated microorganism (rhizosphere). This influences growth by increasing the supply or availability of primary nutrients to the plant. They also play a very important role in decomposition of organic matter, production of plant hormones, and enhancing enzyme synthesis within plants. In conclusion, microorganism contained in controlled-release fertilizers (CRF) has specific functions and effect on plant health and soil functions (Table 13.1).

The microbes used for biofertilizer should be:

- 100% organic, highly complex microorganisms,
- Stimulating plant growth with increased soil microbial balance properties,
- Naturally occurring, harmless, and can safely be used in agriculture,
- Confirmed by DNA identification techniques as active ingredients within the polymicrobial blend,
- Environmentally friendly soil ameliorants.

5. Controlled release of organic fertilizers

Granulated fertilizers that release nutrients gradually into the soil are called controlled-release fertilizer (CRF). Controlled-release particulate fertilizers are formulated with the aim of constant supply of three elements that form the primary nutrient sources necessary for plant growth: N, P, and K. These fertilizers require different release rates to the soil for optimum utilization. N should be released slowly and steadily to meet the growth needs of the plant. On the other hand, K and P do not disperse readily in most soils and they are generally desirable to be released quickly. In particular, P is frequently agronomically undesirable if the nutrient is released slowly [19]. Virtually, all commercially available encapsulated slow release fertilizers containing the three essential nutrients specifically enclose P and K in the core at some fixed ratio. Moreover, a product coated with P and K in the core loses the flexibility in this adjustment hindering the release rate of N necessarily thereby adjusts the release rate of all nutrients to the same level. Here, the difference is in the composition of essential elements directly affecting the controlled release rate. Henceforth, diversity of controlled release biofertilizer formulation can be constructed.

Many fertilizer products are now available in which one or more of the plant nutrients can be released in controlled fashion. This gradual release is determined by the solubility of compounds in the soil moisture. CRF are not water soluble and their nutrients disperse into soil more slowly compared to conventional fertilizers that are soluble in water, whose nutrients are easily dispersed as the fertilizer dissolves. This is because the fertilizer granules may have a semipermeable jacket or an insoluble substrate that prevents dissolution while allowing nutrients to flow outward.

Table 13.1 Categories of Biofertilizers with some examples.

S. No.	Groups	Examples
Nitrogen fixing biofertilizers [17]		
1.	Free-living	*Azotobacter, Beijerinckia, Clostridium, Klebsiella, Anabaena, Nostoc*
2.	Symbiotic	*Rhizobium, Frankia, Anabaena azollae*
3.	Associative symbiotic	*Azospirillum*
Phosphorus solubilizing biofertilizers		
1.	Bacteria	*Bacillus megaterium* var. *phosphaticum*, *Bacillus subtilis* *Bacillus circulans, Pseudomonas striata*
2.	Fungi	*Penicillium* sp., *Aspergillus awamori*, *Aspergillus fumigatus, Aspergillus Niger*
Phosphorus mobilizing biofertilizers [18]		
1.	Arbuscular mycorrhiza	*Glomus* sp., *Gigaspora* sp., *Acaulospora* sp. *Scutellospora* sp. and *Sclerocystis* sp.
2.	Ectomycorrhiza	*Laccaria* sp., *Pisolithus* sp., *Boletus* sp., *Amanita* sp.
3.	Ericoid mycorrhizae	*Pezizella ericae*
4.	Orchid mycorrhiza	*Rhizoctonia solani*
Biofertilizers for micronutrients		
1.	Silicate and zinc solubilizers	*Bacillus* sp.
Plant growth promoting rhizobacteria [14]		
1.	Pseudomonas	*Pseudomonas fluorescens*

5.1 Glass matrix fertilizer (GMF)

A glass matrix fertilizer (GMF) is a product from ceramic industries which has the ability to releases nutrients on the basis of plant demand but only in the presence of metal complexing solutions, a process similar to those exuded by plant roots. This ensures a stable release of nutrients over time, limiting the risk of their loss in the environment. As a procedure to improve fertilizer performance, GMF was mixed with different organic biomasses, such as leather meal, digested vine vinasse, pastazzo (a byproduct of the citrus processing industry PAS), or green compost (COMP).

Organic biomasses mixed with GMF increased the release of macro and micronutrients through an "activation effect," which suggests the application of these organomineral fertilizers for sustainable crops production [20]. For example, glass fertilizers (GF) used for the production of tomato plant were prepared by using a ratio of 0.65 of P_2O_5/K_2O and different concentrations of CaO (0–5.1 wt%), SiO_2 (6.5–56.0 wt%), and Al_2O_3 (0–14.6 wt%). The result reported that the glass fertilizer took more than 40 days for complete leaching out resulting in a prolonged period of biofertilizer release. The harvested tomato fruits and leaves showed similarity with the treatment that received normal applied fertilizer. An increase in crop yield was also noted for glass fertilizer treatments. However, there was no significant difference in soil characteristics [21].

5.2 Organo mineral fertilizer

Nutrient imbalance in agriculture resources can be overcome with the use of biosolids, a waste material commonly known as sewage sludge. Sewage sludge is generally rich in phosphorous but low in potassium and nitrogen. Mineral composition of sewage sludge can be enriched with the addition of urea and muriate of potash as sources of K and N. This results in the production of organo mineral fertilizers having a balanced nutrient requirement for crops. The nutrient balanced sludge-based organo mineral fertilizer is produced by baking digested sewage sludge cake at 80°C in a tumbling evaporator. The sludge granules thus produced are 3–6 mm in diameter. Analyses were carried out based on the crop yield, N use efficiency, and soil fertility. The result showed that the new organofertilizer is as efficient as conventional fertilizer. Moreover, the level of heavy metal in soil was not exceeded than normal. The novelty of this research depends on the ability to transform a waste product into a practical CRF. Hence, it is concluded that organo fertilizer is a promising alternative product for sustainable agriculture [22].

5.3 Sodium bentonite and alginate (NaAlg) composites

Raoultella planticola Rs-2, a plant growth-promoting bacterium, was innovatively encapsulated with various blends of alginate (NaAlg), starch, and bentonite composites to develop efficient slow-release biofertilizer formulations. Sodium bentonite and alginate (NaAlg) composites could be developed to form an efficient slow-release biofertilizer formulation, which minimizes production costs. These formulated microcapsules were spherical in shape with a diameter of dry beads ranging from 0.98 to 1.41 mm having encapsulation efficiency of nearly 100% [23]. These controlled-release formulations had increased biodegradability, swelling, and release rate with increasing NaAlg content and decreased biodegradability with increasing bentonite content. The release kinetics of viable cells from capsules and the swelling ratio of capsules are greatly affected by moisture, temperature, pH, and salt content of the release medium. This work indicated that bentonite—NaAlg-starch (controlled-release formulations of Rs-2 biofertilizer) composites could be an efficient option for low-cost encapsulated wall materials for slow-release bacterial fertilizers in farmlands and have a promising application in natural field conditions [24].

5.4 Lignin-based controlled-release coatings

Urea is the most commonly used fertilizer as a source of nitrogen. Due to its high water solubility, a misuse can easily lead to excess of nitrogen concentration in soil. Lignin was used as an economically feasible and biodegradable slow-release coating for urea CRF [25]. The chief commercially available lignins used are two lignosulfonates (Borresperce and Wafex P), a softwood kraft (Indulin AT), and soda flax lignin (Bioplast). Among these, the latter showed the best potential with respect to film-forming properties. The dry matter content of Bioplast dispersions is processable up to 50% but sometimes results in thin coating layers due to high losses during processing. To minimize this hazard during urea release, hydrophobic compounds and cross-linkers were added to the Bioplast dispersions. In addition, alkenyl succinic anhydride (ASA) significantly reduces the release of urea in water. The time required for the complete urea release on an average is 1 hour, and thus depends on a low reactivity of the selected compounds toward lignin, negative effects of the selected compounds, or low percentages of applied coating during film-forming process. Since urea is highly water-soluble, it partly dissolves in the aqueous lignin dispersions, resulting in low water resistance coatings. To overcome this, an inner coating layer with high dry matter content is applied. In conclusion, for industrial application lignin shows high potential as coating material and hence the film-forming properties are desired.

5.5 Hydrophilic polymers

Using modern technology diverse packaging and delivery devices have been used to supply nutrients to plants at a controlled rate in the soil. One new innovative approach used today is hydrophilic polymers as carriers of plant nutrients.

These polymers are generally classified as

(1) Synthetic polymers: derived from polymerization of monomers from petrochemicals.

(2) Semisynthetic polymers: based primarily on cellulose (such as wood pulp), which is reacted with functional groups derived from petrochemicals.

(3) Natural polymers: including proteins, polysaccharides, lignins, and rubber (derived from polysaccharides).

By controlling various reactions condition during polymer formation, various degrees of cross-linking, anionic charge, and cationic charge can be added and thereby, change their effectiveness as fertilizer. With respect to conventional fertilizer, when fertilizer containing solution is mixed with hydrophilic polymer to form a gel, the release of soluble nutrients is delayed. The gel has to be formed prior to application in soil. The factors that determine the effectiveness of a controlled-release hydrophilic polymeric system are physical and chemical properties, its biodegradation rate, and the fertilizer source used. In some circumstances addition of polymer with nutrients has been shown to reduce N and K leaching from well-drained soils and also enables the plant to recover N, P, Fe, and Mn from the soil [26].

5.6 Macromolecular slow-release fertilizer (MSF)

The technology of macromolecular slow-release fertilizer (MSF) containing N, P, and K plays an important role in improving fertilizer use efficiency by plants and also reduces the frequency of fertilization, thereby mitigating environmental pollution and leading to the development of sustainable agriculture. The chief components of macromolecular fertilizer are nitrogen, phosphorus, and potassium. The structural analysis was characterized by the FTIR spectrum and the gel permeation chromatography showing the average molecular weight of the fertilizer to be 13,500. Moreover, the effect of the decomposition of the macromolecular fertilizer in soil and water is environmentally safe [27].

5.7 Coffee spent grounds (CSG)

A sustainable fertilization technique which might be even performed by unqualified individuals is contributed by mixing mineral material with organic coffee spent grounds (CSG). CSG are formed as a byproduct after the preparation of coffee (under pressure of 15 atm). The spent grounds were initially dried in sun and then dried to a constant mass at a temperature of 105°C. The compound gelatin was used as a binder, which is prepared by soaking in deionized water and later boiling it until it forms a gel. The mineral supplement was made of ash formed during the incineration of oak wood. The oak wood in the form of woodchips along with bark was incinerated at a temperature of 600°C for 3 h in the muffle furnace. The capsules are covered by using cellulose sheets and collagen stick. Slow decomposition of the tablets formed gradually releases elements to the environment which prolongs the time of the fertilizer's influence. Thus, the production of fertilizer tablets from CSG, biomass ash, and magnesium sulfate not only minimizes the mass of produced wastes, but also contributes to improving the productivity of soils [1].

6. Conclusion

Organic farming primarily depends on the cycling of organic matter to the soil to maintain fertility. Organic fertilizers that are free or substantially free from toxic chemicals or metals can be used as a component of fertilizers either alone or in combination with an inorganic fertilizer. Compost, plant byproducts, cover crops, animal manure, and other biological materials are applied to organic fields to improve yield. In other ways, leftovers of organic waste or plant and animal byproducts from primary industry such as blood, fish emulsion and bone meal, cottonseed meal, wood ash are all classified as organic fertilizers. Organic fertilizer increases the organic composition of the soil along with the major and minor organic nutrients. The natural products, such as mined minerals are also permitted to supplement organic matter to the soil. Similar to inorganic fertilizers, organic fertilizers may be natural or synthetic. The factor that differentiates both is the carbon, more specifically, the carbon–hydrogen linkage in organic fertilizers, which slows the release of the nutrient ions. The slower nutrient release improves overall physical characteristics of the soil, thus improving sustained availability of the nutrients, soil aeration, and prevents root burn and leaching losses compared to inorganic fertilizers. Organic fertilizers act as an energy source for microorganisms in the soil, thereby improving soil structure and plant growth. They also release nitrogen, phosphate, and potash in a manner easily absorbed by plants. Organic fertilizers provide a long-term protection to agricultural land from harmful effect of inorganic fertilizers, which leads to soil damage and pollution.

References

[1] T. Ciesielczuk, C. Rosik-Dulewska, E. Wiśniewska, Possibilities of coffee spent ground use as a slow action organo-mineral fertilizer, Annu. Set the Environ. Prot. 17 (2015) 422—437.

[2] S. Niassy, K. Diarra, Y. Niang, S. Niang, H.R. Pfeifer, Effect of organic fertilizers on the susceptibility of tomato lycopersiconesculentum: solanaceae to Helicoverpaarmigera Lepidoptera: noctuidae in the niayes area Senegal, Res. J. Agric. Biol. Sci. 6 (6) (2010) 708—712.

[3] A.M. Khomami, I. Guilan, The possibility using the composted peanut shells in the growth .of marigold and Viola tricolor plants, J Ornam. Plants 5 (1) (2015) 61—66.

[4] S.N. Dauda, F.A. Ajayi, E. Ndor, Growth and yield of water melon (Citrulluslanatus) as affected by poultry manure application, J. Agric. Soc. Sci. 4 (3) (2008) 121—124.

[5] S.J. Naylor, R.D. Moccia, G.M. Durant, The chemical composition of settleable solid fish waste (manure) from commercial rainbow trout farms in Ontario, Canada, N. Am. J. Aquacult. 61 (1) (1999) 21—26.

[6] J.K. Whalen, C. Chang, G.W. Clayton, J.P. Carefoot, Cattle manure amendments can increase the pH of acid soils, Soil Sci. Soc. Am. J. 64 (3) (2000) 962—966.

[7] Z.A. Stephan, The Efficacy of Nematicides and Horse Manure in Controlling Root-Knot Nematodes on Tomato and Eggplant, Nematologia Mediterranea, (Italy), 1995.

[8] J.L. Walworth, C.R. Woolard, K.C. Harris, Nutrient amendments for contaminated peri-glacial soils: use of cod bone meal as a controlled release nutrient source, Cold Reg. Sci. Technol. 37 (2) (2003) 81—88.

[9] L. Chen, J. Kivelä, J. Helenius, A. Kangas, Meat Bone Meal as Fertiliser for Barley and Oat, 2011.

[10] J. Spångberg, P.A. Hansson, P. Tidåker, H. Jönsson, Environmental impact of meat meal fertilizer vs. chemical fertilizer, Resour. Conserv. Recycl. 55 (11) (2011) 1078—1086.

[11] B. Gagnon, S. Berrouard, Effects of several organic fertilizers on growth of greenhouse tomato transplants, Can. J. Plant Sci. 74 (1) (1994) 167—168.

[12] C. Ciavatta, M. Govi, L. Sitti, C. Gessa, Influence of blood meal organic fertilizer on soil organic matter: a laboratory study, J. Plant Nutr. 20 (11) (1997) 1573—1591.

[13] S. Sivasankari, V. Venkatesalu, M. Anantharaj, M. Chandrasekaran, Effect of seaweed extracts on the growth and biochemical constituents of Vigna sinensis, Bioresour. Technol. 97 (14) (2006) 1745—1751.

[14] J.K. Vessey, Plant growth promoting rhizobacteria as biofertilizers, Plant Soil 255 (2) (2003) 571—586.

[15] L. Herrmann, D. Lesueur, Challenges of formulation and quality of biofertilizers for successful inoculation, Appl. Microbiol. Biotechnol. 97 (20) (2013) 8859—8873.

[16] M. Atieno, L. Herrmann, R. Okalebo, D. Lesueur, Efficiency of different formulations of Bradyrhizobium japonicum and effect of co-inoculation of Bacillus subtilis with two different strains of Bradyrhizobium japonicum, World J. Microbiol. Biotechnol. 28 (7) (2012) 2541—2550.

[17] V. Devi, V.J.H. Sumathy, Biofertilizer Production from Agro-Wastes, 2017.

[18] A.A. Khan, G. Jilani, M.S. Akhtar, S.M.S. Naqvi, M. Rasheed, Phosphorus solubilizing bacteria: occurrence, mechanisms and their role in crop production, J agric biol sci 1 (1) (2009) 48—58.

[19] K.E. Fersch, W.E. Stearns, U.S. Patent No. 4,042, U.S. Patent and Trademark Office, Washington, DC, 1977, p. 366.

[20] A. Trinchera, M. Allegra, E. Rea, G. Roccuzzo, S. Rinaldi, P. Sequi, F. Intrigliolo, Organo-mineral fertilisers from glass-matrix and organic biomasses: a new way to release nutrients. A novel approach to fertilisation based on plant demand, J. Sci. Food Agric. 91 (13) (2011) 2386—2393.

[21] A. Tamayo, R. de la Torre, O. Ruiz, P. Lozano, M.A. Mazo, J. Rubio, Application of a glass fertilizer in sustainable tomato plant crops, J. Sci. Food Agric. 98 (12) (2018) 4625–4633.

[22] L.K. Deeks, K. Chaney, C. Murray, R. Sakrabani, S. Gedara, M.S. Le, G.H. Smith, A new sludge-derived organo-mineral fertilizer gives similar crop yields as conventional fertilizers, Agron. Sustain. Dev. 33 (3) (2013) 539–549.

[23] Y. He, Z. Wu, L. Tu, Y. Han, G. Zhang, C. Li, Encapsulation and characterization of slow-release microbial fertilizer from the composites of bentonite and alginate, Appl. Clay Sci. 109 (2015) 68–75.

[24] Z. Wu, L. Guo, S. Qin, C. Li, Encapsulation of Raoultella planticola Rs-2 from alginate-starch-bentonite and its controlled release and swelling behavior under simulated soil conditions, J. Ind. Microbiol. Biotechnol. 39 (2) (2012) 317–327.

[25] W.J. Mulder, R.J.A. Gosselink, M.H. Vingerhoeds, P.F.H. Harmsen, D. Eastham, Lignin based controlled release coatings, Ind. Crop. Prod. 34 (1) (2011) 915–920.

[26] R.L. Mikkelsen, Using hydrophilic polymers to control nutrient release, Fert. Res. 38 (1) (1994) 53–59.

[27] G.Z. Zhao, Y.Q. Liu, Y. Tian, Y.Y. Sun, Y. Cao, Preparation and properties of macromolecular slow-release fertilizer containing nitrogen, phosphorus and potassium, J. Polym. Res. 17 (1) (2010) 119–125.

Index

Note: Page numbers followed by "t" indicate tables, "f" indicate figures and "b" indicate boxes.

Printed in the United States
By Bookmasters